Forest Litter Insect Communities
Biology and Chemical Ecology

Forest Litter Insect Communities

Biology and Chemical Ecology

T.N. ANANTHAKRISHNAN

SCIENCE PUBLISHERS, INC.
10 Water Street, #310, Lebanon, NH 03766, USA

Science Publishers, Inc.
10 Water Street, # 310
Lebanon, NH 03766, USA

ISBN 1-886106-58-4

Printed at Baba Barkha Nath Printers, New Delhi.

Dedicated to Menaka

Preface

The role of litter as a 'transitory bank of nutrients' is well recognized besides being a factor affecting arthropod communities and their dynamics, more especially through alteration of the physical and chemical environment. The complexity of bacterial, fungal and faunal community structure, their interrelationships and function in organic matter decomposition and nutrient cycles are essential aspects of litter studies. This is especially so in the decomposing litter of Tropical Forests where a knowledge of the community studies becomes important and relevant. Community structure of organisms in litter depends on precise information on biodiversity of the forest floor, since the microbial and faunal elements play a major role in organic matter turnover. While fungi are the most important primary colonizers of fresh litter, there is also a succession of fungi during decomposition process. Vertical movements up and down the organic horizon introduce bacterial and fungal inocula from the lower organic layer to the freshly fallen litter. Food webs and trophic levels become complicated because of the large number of interactions of the components. Movement of matter through the food web largely determines soil fertility.

Plants contribute forest organic material to surface of the soil. The forest litter is speedily attacked by microbial and faunal elements which during the process of decomposition, rapid under aerobic conditions with adequate moisture ultimately get converted into the heterogenous mixture called humus. The undisturbed forest ecosystem also operates in a closed nutrient cycle, since the nutrients released by decomposition of soil organic matter are quickly absorbed. The relative roles of Ca, K, Na and Mg in the decomposition process not only in the different layers of litter, but also on the quality of litter layer at mean sea level and at varying altitudes, are also important aspects. Of added interest is the role of secondary chemicals notably phenolics, flavonoids, tannins and diverse volatiles released from litter which have considerable effect on faunal diversity.

This work aims at a synthesis of available knowledge of litter ecosystem in Tropical Forests, from the viewpoint of the sustainability of entomological resources at various levels, laying stress on their biological diversity with emphasis on their functional diversity. At a time when we are losing natural forests at a fast rate through conversion of forest into cultivated land, the thoughts of dwindling biodiversity through lack of litter become predominant. Of equal relevance is the conversion of natural forest into monoculture plantations such as teak, reducing litter diversity and consequently species diversity. Information presented in this volume which highlights aspects of decomposition of litter and mineral cycling, lays emphasis on the biodiversity of biotic elements involving: (a) microbial–faunal interactions, (b) bioecology of diverse faunal elements, and (c) predator dynamics. Each of these aspects refer not only to monoculture and natural forest litter, but also to the degree of diversity at various altitudes in relation to the quality of litter resource. Trophic systems in forest litter communities have also been discussed, so that emphasis is made on community structure in litter ecosystems. Chemical ecological studies of litter involving primary and secondary metabolites as well as leaf volatiles now known to play an important role in the structure of litter communities, have been discussed at length.

This work was undertaken during the period of the Pitambar Pant Award of the Ministry of Environment and forests, Government of India, New Delhi. In the preparation of this book, I have received help from diverse sources, in particular from Dr. R. Gopichandran, Dr. Sabu K. Thomas and Mr. A. Jeyakumar, who were associated with me as Research Assistants during different periods. The success of this venture is also due to the large-heartedness of Dr. K.R. Seshan of the Texas A & M University, Austin, for his consistent support with literature on various aspects of Forest Ecology. I also wish to thank the Zoological Survey of India, Calcutta, in particular Drs. A.K. Hazra, K. Sanyal and Biswas, for identification of insects sent periodically. To Dr. K. Natarajan of the Botany Department, University of Madras, my thanks are due in no small measure for his help in the identification of fungi.

T.N. ANANTHAKRISHNAN

Madras
10.9.95

Contents

1

Litter Dynamics and Role of Litter in Forest Ecology

Introduction

Litterfall forms one of the facets of nutrient cycling in forests, representing a major process for the transfer of nutrients from aboveground vegetation to soils. Litter includes the layer of dead plant material present on the soil surface and comprises the surface litter (L-layer), the partially decomposing layer beneath it (F-layer) and the humus layer (H). Litter production depends primarily on the productivity of the plant community at the site, and exhibits seasonal patterns varying with vegetation type and latitude, besides altering the physical and chemical environment. Chemical and physical degradation, heterotrophic consumption and decomposition reduce litter accumulation on the surface. These are interrelated processes with decomposition being the most important (Facelli and Pickett 1991). Litter plays a major role in the transfer of energy and nutrients in the forest ecosystem and litterfall data have been used to predict the productivity of the ecosystem (Toky and Singh 1993).

Polyphenols, organic acids and other substances percolate from the surface litter during litter decomposition and the quality of the litter layer is an equally important element, so that litter is considered as a biological system in which organic matter, mineral component, microflora, soil fauna and vegetation interact. The breakdown of the litter and organic material release nutrients into forms available to plants and micro-organisms enabling completion of the nutrient cycle in a forest ecosystem. The soils are well drained, well aerated, but once the forest disappears they are susceptible to leaching. Hence the importance of litter which often includes leaves, dropped fruits, dead twigs and bark and the quality of litter naturally varies with the age of the trees, with climate as well as the nature of vegetation with a mixed profile, and diversity in species composition, very evident in natural forests as against man-made monoculture forests.

The topography of the forests also plays a part in relation to the nature of litterfall, such as whether it is a montane tropical forest or a lowland forest. The fine litter type, involving branches and falling tree trunk is more heterogenous both in space and time. Tropical lowland evergreen rain forests occur where water stress is absent; Tropical semi-evergreen rain forest is a closed high forest with very tall emergent trees. The upper montane rain forest supports trees around 10 m tall, unlike the semi-evergreen where canopy height reaches 45 m. The tropical lower montane forest is intermediate with a canopy height of 15–33 m. Tropical species being more habitat specific, have narrow tolerance ranges to physical conditions and this provides opportunities for altitudinal studies of tropical litter insects, which tend to show a decline in abundance and diversity at higher elevations and a mid-elevation peak in species richness (Olson 1994). High tree diversity could result in a greater variety of litter sources and conditions for many litter faunal elements. The uniformly moist and warm conditions at mid-elevations tend to be conducive to decomposer and detritivore activity.

Patterns of Litterfall

The patterns of litterfall tend to vary with climate (Bray and Gorham 1964; Rodin and Basilewich 1967; Klinge 1978) and exposure. Data concerning litterfall is available for 40 tropical forests around the world, encompassing leaf fall, fruits, and twigs. Litterfall serves to estimate the production dynamics of the rapid growing and highly dynamic part of the vegetation. Litter averaged 10 t/ha/a in these 40 forests and the frequency distribution of litter indicates that there may be 4 groups of data representing very dry forests, dry forests, semi-deciduous forests and wet forests. Highest litterfalls were 3–25 t/ha/a, leaf litter contributing to 54% of the total litter and in 2,100 m tropical humid montane forests of South India, leaf litter constitutes 70%. The average total litter for 3 years amounted to 5.6 t/ha/a of which leaf litter represents 3.9 t/ha/a and 1/3 t/ha/a contributed by twigs and bark (Blasco and Tanny 1975). In monoculture forests like teak (*Tectona grandis*) in northern India, leaf litter was 5 t/ha/a, in *Terminalia-Shorea* forest (6.2%), *Diospyros-Anogeissus* (4.2%), *Shorea-Buchanania* (3%), *Butea* (1%) and in dry deciduous forests it is 6.2 for leaves and 9.3% for stem and branches and montane forests at 2,100 m (3%) (UNESCO 1978). The massive defoliation occurring in some species explains the high increase in leaf litter production in some years. Blasco and Tanny (1975) measured litter production for over 3 years and found that it varied from 3.8 to 7.3 t/ha/a averaging 5.5 in Tropical, humid montane forests of South India.

In equatorial forests the production of litter components has been shown as 9.5 t/ha/a (leaves), 19.2 t/ha/a (stem and branches) and 2.8 t/ha/a of roots. Information relating the litterfall patterns of important species has been documented by Pascal (1980) in the Attapadi forests of Western Ghats wherein the annual litter production showed a mean value of 8.5 t/ha/a (with leaves and foliar fragments contributing 6.4 t) as compared to 3.4 t/ha/a for the montane forests of Gundar of South India (Blasco and Tanny 1975). Attapadi, though a pure evergreen forest had similar leaf shed rhythms as the other deciduous forests, heavy litterfall occurring during the dry season, the maximum being during January to March with peak in February.

Phenology of Litter Production

As for the phenology of litter production, interestingly enough, the leaf shedding patterns differ according to the species. For some, like *Persea macrantha* and *Prunus ceylonica* the leaf shedding takes place form January to February, in *Cassia glauca* from January to March, in *Drypetes elata* and *Turpinia malabarica* from December to January. For others like *Messua ferrea*, the leaf-fall was massive in only one year, while others like *Aglaia anamallayana* and *Calophyllum polyanthum* have two periods of leaf-fall in both dry and wet seasons. *Cullenia exarillata* and *Palaquium ellipticum* account for 49.2% of the forest litter (Pascal 1980) (Fig. 1). Attapadi being an evergreen climate forest, interannual variations in ground litter correspond to climatic variations. The mean total value of litter was 13 t/ha/a composed of leaves, (32.6%), wood (36.4%) and debris (25.4%), the total quantity varies between 11.89 and 15.09 t/ha. The quantity of wholesome leaf litter in the forest floor varied between 2.4 t/ha and 6.7 t/ha/a, and there is an accumulation from December to March followed by a plateau or gradual decrease. The decrease accelerated between September and December towards the end of the rainy season corresponding to a revival of biotic activity in the soil (Pascal 1980) (Fig. 2). That litterfall patterns are complex and vary with the type of vegetation is further substantiated by observations recorded in monoculture stands of *Tectona grandis*, wherein litter production varied from 6.1 to 17.3 t/ha/a (Singh 1968), with maximum leaf-fall from February to July prior to the onset of monsoon. Litter production in teak, sal (*Shorea robusta*), eucalyptus and pine under similar macroclimatic conditions for Dehra Dun is known

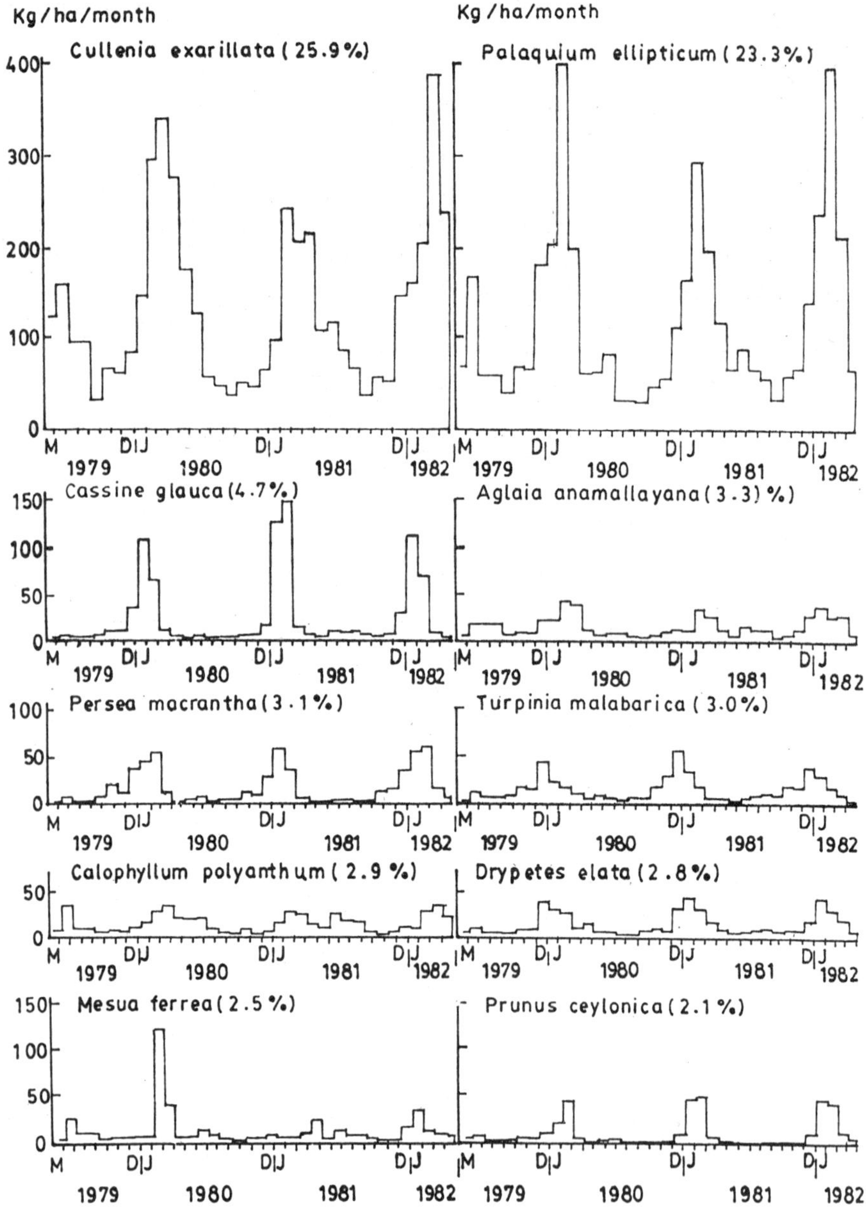

Fig. 1 Phenology of litter production in the Attapadi forest (after Pascal, 1980).

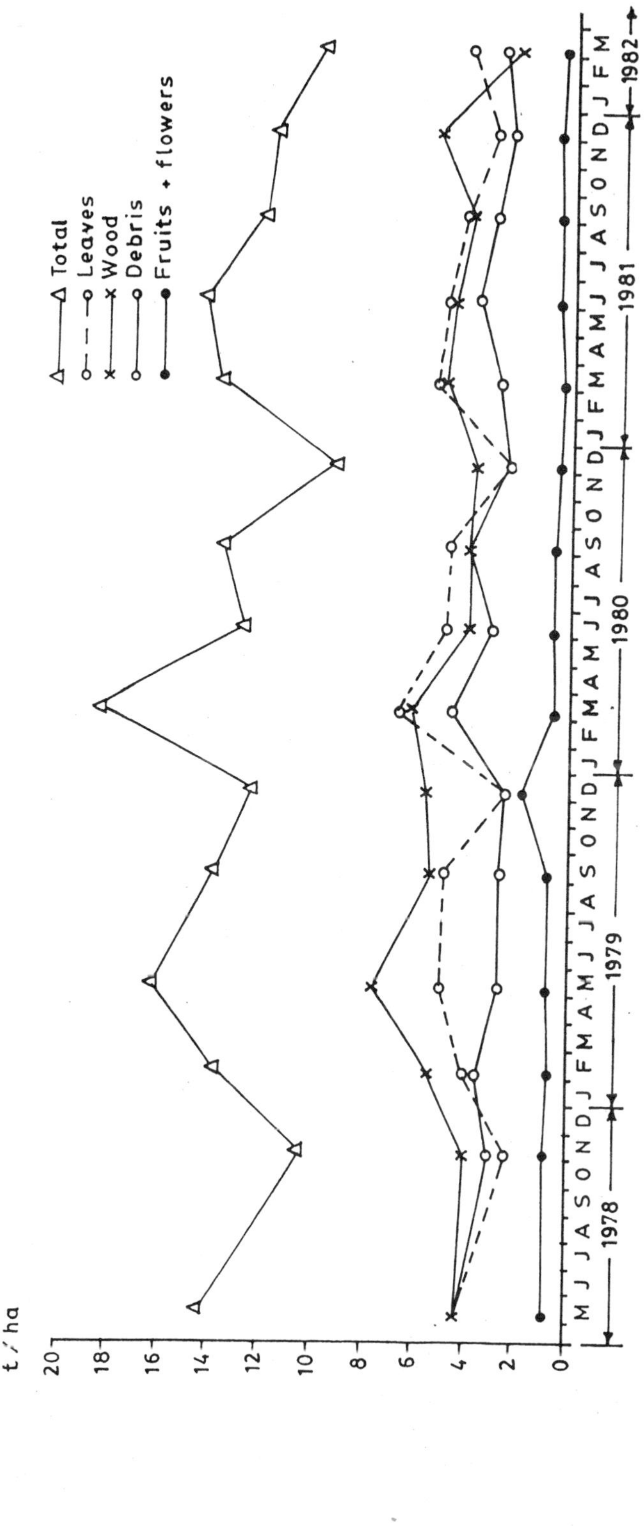

Fig. 2 Attapadi-variations in the forest floor litter (after Pascal 1980).

to be of the order of 6.95, 11.27, 9.67 and 7.06 t/ha/a (Pande and Sharma 1993) (Figs. 3, 4).

Production of litter in 4-year-old stands of *Leucana latisiliqua* (Syn. *L. leucocephala*) and *Eucalyptus tereticornis* studied by Toky and Singh (1993) indicates similarity, with about 7.5 t/ha/a and 7.2 t/ha/a and twice the production in *Acacia nilotica*. But the quantity of litter in 8-year-old stands was 11.5 t/ha in *L. latisiliqua*, 9.1 t/ha in *E. tereticornis* and 5.4 t/ha in *A. nilotica* (Table 1). All the stands showed a seasonal trend in litterfall producing maximum during summer and minimum during winter, resulting in bimodal peaks in annual pattern in the litterfall. The high rate of litterfall during summer probably results from high temperature and water scarcity (Fig. 5).

In evergreen forests, the stands being floristically heterogenous, litter contribution by different species partly explain the variation in nutrient concentration in different seasons. The humid environment with a dense forest cover, low temperature, higher moisture content and relatively undisturbed nature of the forest, support a luxuriant growth of several species resulting in greater production of litter (Ananthakrishnan and Sabu Thomas 1993; Visalakshi 1993).

With climate as the basis of classification, the major forest categories contribute to the tropical, subtropical, mediterranean and warm/cold temperate types, while they are characterised as either broad-leaved or narrow-leaved based on the basis of life form. Behavioural patterns as phenological events determine their classification as either evergreen or deciduous (Holdridge *et al.* 1971). The average litterfall on a world-wide scale is summarised in Table 2 (Vogt *et al.* 1986).

Interestingly, however, comparable quantities of litterfall close to 100,000 kg/ha have also been recorded in contrasting cold temperate, broad-leaved and needle-leaved deciduous, semi-deciduous forest types and the evergreen forests, while in tropical and subtropical types accumulation has also been reported at less than 2,500 kg/ha. Accordingly, correlations of climatic factors with organic matter accumulation cannot be generalised, which also indicates that litter accumulation may vary by several magnitudes even in forests under similar climatic zones (Meentemeyer 1978; Meentemeyer *et al.* 1982).

The importance of tropical forests to the dynamic processes of the biosphere is very significant and tropical forests have greater gross production and leaf litterfall is an index to forest productivity, even though tropical forests produce less wood per leaf mass than temperate forests (UNESCO 1978). Litter turnover in tropical forests is calculated as annual litterfall (kg/ha) divided by the litter standing crop on the soil surface. The annual litterfall for a forest in equilibrium is generally assumed to equal its decomposition and for this the average amount

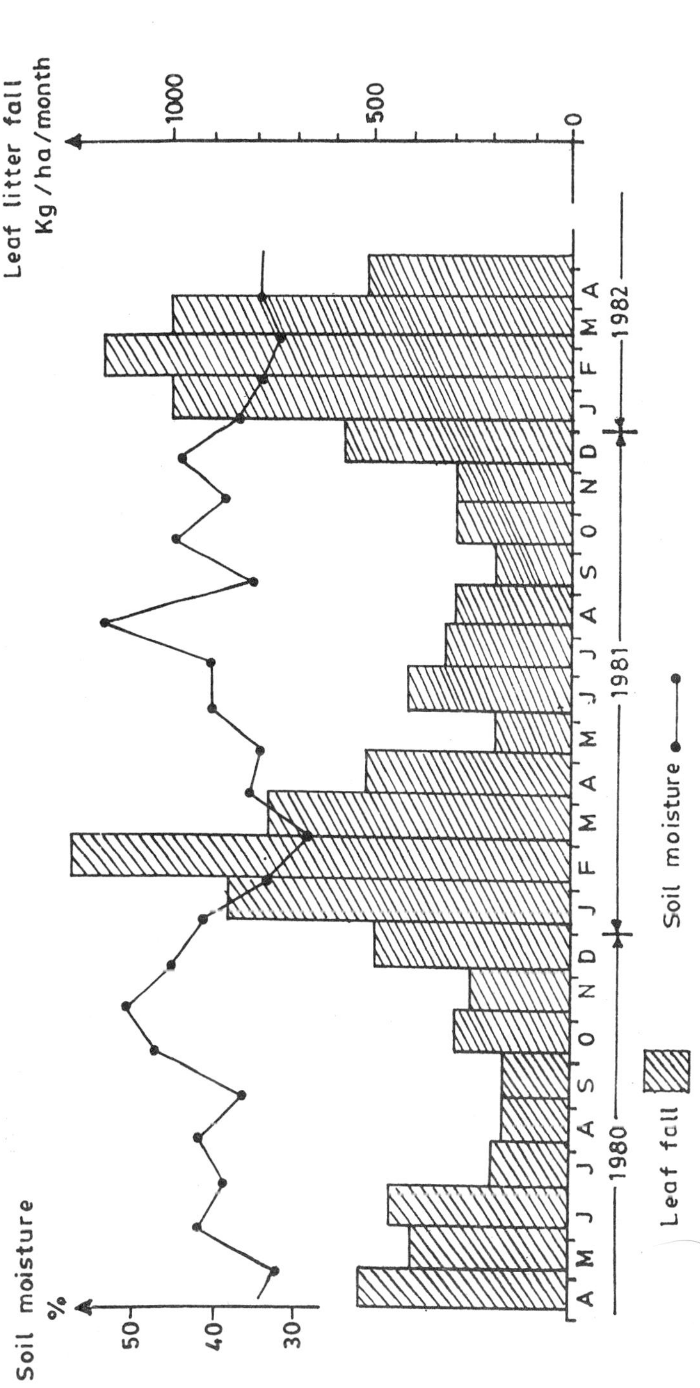

Fig. 3 Attapadi-monthly variations in the leaf-fall, soil moisture and saturation deficit (after Pascal, 1980).

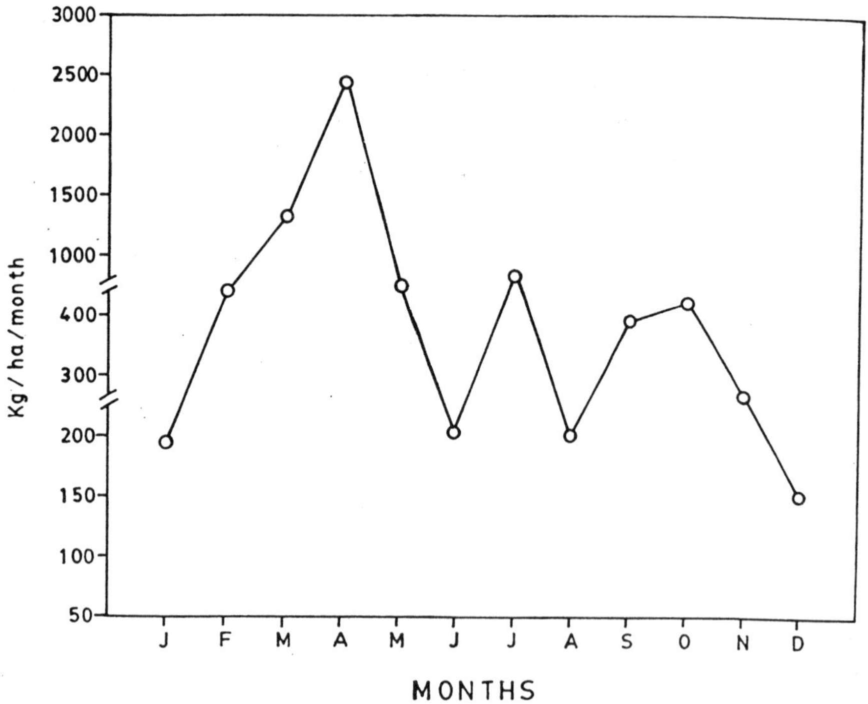

Fig. 4 Patterns of litterfall in *Tectona grandis* monocultures.

Table 1. Annual litterfall (t/ha/a) in 4- and 8-year-old energy plantations of *A. nilotica*, *E. tereticornis* and *L. latisiliqua*. Values in parentheses are percentages of the total (after Toky and Singh 1993).

Species	Litter component			Total
	Leaves	Fruits	Twigs	
4-year-old				
A. nilotica	3.07 ± 0.140 (72%)	0.96 ± 0.210 (23%)	0.19–0.010 (5%)	4.22± 0.295
E. tereticornis	4.51 ± 0.141 (62%)	0.25 ± 0.213 (3%)	1.89 ± 0.162 (26%)	6.85 ± 0.255
L. latisiliqua	4.99 ± 0.294 (66%)	1.58 ± 0.249 (21%)	0.95 ± 0.090 (13%)	7.52 ± 0.318
8-year-old				
A. nilotica	3.71 ± 0.124 (69%)	1.34 ± 0.268 (25%)	0.31 ± 0.012 (6%)	5.36 ± 0.292
E. tereticornis	5.32 ± 0.160 (58%)	0.81 ± 0.036 (9%)	1.88 ± 0.128 (21%)	8.01 ± 0.412
L. latisiliqua	89.62 ± 0.417 (72%)	2.37 ± 0.243 (20%)	0.95 ± 0.054 (8%)	11.94 ± 0.648

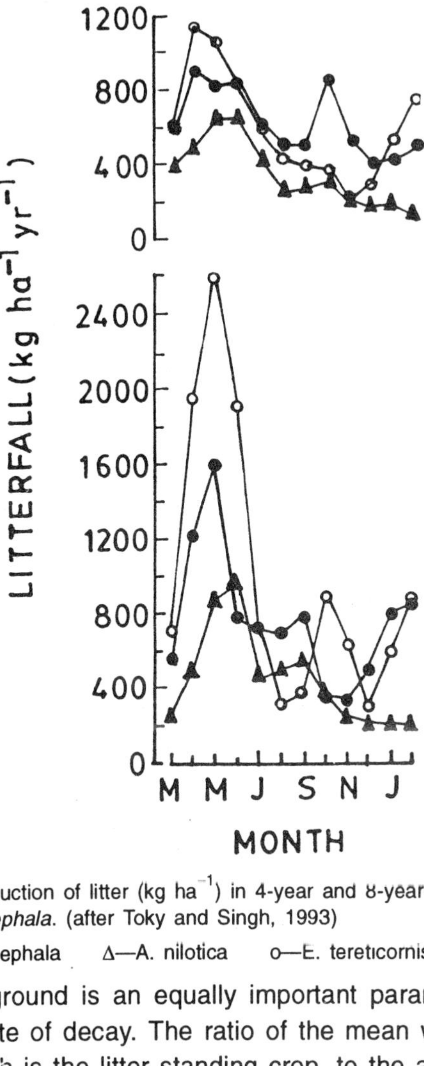

Fig. 5 Monthly production of litter (kg ha^{-1}) in 4-year and 8-year-old energy plantation of *L. leucocephala*. (after Toky and Singh, 1993)

• —L. leucocephala   Δ—A. nilotica   o—E. tereticornis

of litter on the ground is an equally important parameter in order to determine the rate of decay. The ratio of the mean weight of litter on the ground, which is the litter standing crop, to the annual litterfall for the average time in years, is a useful index for litter decomposition. Tropical rain forests with plant and animal communities having a rich diversity of species and a complex web of interactions are likely to be more fragile than relatively simple and robust temperate ecosystems. The reproductive mechanisms of tropical rain forests plant and animal species are more adapted to biological competition than large-scale environmental disturbances. Tropical ecosystems tend to store more nutrients in the vegetation with differential distribution also in the litter

Table 2. The average litter fall on a world-wide scale.

Forest type	Litterfall kg/ha/a
Tropical broad-leaved deciduous	8,438
Tropical broad-leaved evergreen	9,369
Tropical broad-leaved semi-deciduous	5,890
Subtropical broad-leaved deciduous	3,333
Subtropical broad-leaved evergreen	5,098
Warm temperate broad-leaved deciduous	4,236
Warm temperate broad-leaved evergreen	6,484
Warm temperate needle-leaved evergreen	4,432
Cold temperate broad-leaved deciduous	3,854
Cold temperate needle-leaved deciduous	3,590
Cold temperate needle-leaved evergreen	3,144

that constitutes the aboveground biomass than in the soil matrix studies.

Biomass in an undisturbed forest is maximal under the local environmental conditions and reflects the capacity of the ecosystem to accumulate organic matter. An equilibrium between production of new biomass and loss of living biomass due to mortality, is always evident. Litter decomposition is closely related to biogeochemical cycles and the rate of litter decomposition determines the stores of inorganic elements remaining in the litter component. Seasonal changes are equally evident and Tropical Forests differ from most ecosystems in their very high vegetation biomass, the very high elemental stores in this biomass and their rapid rates of recycling. Some of these aspects are discussed in the next chapter of the dynamics of litter decomposition in forest ecosystems.

2

Dynamics of Litter Decomposition

Plant litter is decomposed by a sequence of events involving physical processes such as leaching and mechanical breakdown, as well as biological processes such as microbial degradation involving several exoenzymes (Sinsabaugh and Linkins 1989; Kshattriya *et al.* 1992). The rates of litter decomposition are variable, often correlated with the input levels of nutrients and soil fertility. Montane forests have low temperatures and hence lesser rates of decomposition of litter (Grubs 1977; Tanner 1981 and Coleman *et al.* 1983). The development of bacterial flora after litterfall initiates litter breakdown and due to increased moisture conditions there is considerable increase in bacterial flora. In easily decomposable litter the number may reach very high values in a short time and then decrease gradually. Decomposition of litter is regulated by a host of variables including litter physical properties, climate, macro- and microfaunal responses. The two most important contributors of litter decomposition rates are climate and susceptibility to attack by special decomposers, making substrate quality an important parameter.

Lignin concentration of the substrate is an excellent index to use for predicting the rate of disappearance and weight loss by forest litter samples (Meentemeyer 1978). Phosphorus and calcium concentration, lignin and tannin, carbohydrate including cellulose have significant correlations with decomposition rate. Slowly decomposing litter components such as lignin tend to dominate the shape of the long-term decomposition rate. Due to the ubiquity of cellulose, cellulases are the most studied enzyme systems in plant litter. Unlike cellulose and amylase activity, invertase activity is seen to be highest at the beginning of litter decomposition of all plant species. Invertase activity varies with plant species, being higher in the litter at lower altitudes and is positively correlated and generally significant with litter total nitrogen (Kshattriya *et al.* 1992). The production of microbial enzymes is influenced by the chemical nature of the enzymes. Slow rates of litter decomposition can result in the accumulation of large nutrient stuff in the soil surface horizon. Chemical indices of substrate quality include element concentration and that of various classes of organic compounds. Nitrogen content of plant material is important in controlling the rate of decomposition and lignin content exerts more control over the rate of decomposition than does nitrogen (Mellilo *et al.* 1982). Higher the initial lignin content of the material, larger the change in the nitrogen concentration of the material per unit carbon respired during decomposition. So also higher the initial lignin content, the greater is the amount of the nitrogen immobilised per unit carbon respired. Finally, the relationship between the humus formation and initial lignin content of litter has implications for the rate of nitrogen cycling in forest ecosystems. Free amino acids tend to decrease with advance in humification of litter, and these free amino acids are derived from the amino acid pool of the fresh litter. Free amino acids can also result from degradation of cellular protein and polypeptides as well as from the pool of amino acid formed by natural microbial synthesis during humification. While there is a general tendency for a decrease, such amino acids as proline, arginine and gamma aminobutyric acid show slightly increasing concentrations (Lahdesmaki and Pushpanen 1989).

Role of Microbes in Litter Decomposition

Tropical evergreen forests restricted to areas of lesser nutrient levels in the process of optimization of nutrients in canopy, release only much lesser quantities through litter. This further reduces the nutrient inputs into the soil and consequently retard decay rates due to the resultant reduced microbial activity (Swift *et al.* 1979). In the later stages of decomposition the litter may be invaded also by soil bacteria and filamentous fungi, thus establishing a succession wherein the initial colonisers gradually disappear and are replaced by Basidiomycetes with such species as *Penicillium, Trichoderma, Mucor* etc. It is now well established that fungi are recognised to be the chief colonisers and decomposers and the bacteria are actually the secondary decomposers. In their studies on litter of *Eucalyptus*, teak and *Albizzia*, Mary and Sankaran (1991) have shown that the population of bacteria, fungi and Actinomycetes colonising decomposing litter vary significantly between litter species. *Eucalyptus* litter harboured lesser number of bacteria and fungi than that of *Albizzia* and the Actinomycete population was lesser than in teak. The variations in the number of microbes during different months may also be attributed to differences in climatic conditions and litter moisture content (Sinha and Dayal 1983).

Mary and Sankaran (1991) have isolated 92 genera of fungi from teak litter; the fungi imperfecti represented by 28 genera (87.5%), Zygomycetes by 3 genera (9.4%) and one Ascomycete. *Aspergillus, Penicillium, Coniella granati, Myriothecium verrucaria* and *Chaetomella circinoseta* were the primary colonisers. *Aspergillus flavus, A. niger, Curvularia lunata* and species of *Fusarium, Alternaria* and *Myriothecium* were the secondary saprophytic fungi. During the advanced stages of decomposition, *Mucor* and *Rhizopus* were isolated. Basidiomycetes showed high relative density during advanced stages of decomposition including such forms as *Absidia cylindrospora, Cunninghamella bertholettiae, Bastalinia robillardoides, Collybia leucophaea, Lasiodiplodia theobromae, Poecilomyces* sp., and *Scolecobasidium variabile*. Members of fungi imperfecti are strong colonisers on decomposing litter and the Zygomycetes and Actinomycetes do not readily establish themselves. Of the initial colonizers the most important ones include species of *Curvularia, Cladosporium* and *Alternaria*. Such species as *Pithomyces sacchari, Stenophylium zygoshynctus* appear restricted to eucalyptus litter. Fungal flora of teak, eucalyptus and polyculture litter have shown distinct differences (Ananthakrishnan *et al.* 1983a, b). Species assemblages tend to vary at the different sites, for instance, forest litter of *Tectona grandis* at mean sea level (MSL) and 250 m, was found to harbour *Trichoderma* sp., while at higher altitudes, species of *Aspergillus* and *Fusarium* were isolated. The F layer included a wider range of species at MSL, whereas at 250 m only *Gliocladium* was isolated, similar to a single species dominance at 1,300 m. Similarly, only *Aureobasidium pullans* occurred in the humus layer at MSL. Nutrient-poor litter harbours proportionately lesser populations of microbes since the energy requirements are easily met with. Active participation by microbes in decomposition contributes to spatial transfer of nutrients to sites of progressive degradation of litter, thus significantly altering the chemical environment favouring enhanced rates of metabolic transformations (Vogt *et al.* 1983).

Recent investigations on the dynamics of microbial populations in relation to changes in the chemical ecological characteristics of litter have indicated that litter in polyculture complex forests harbour large populations of bacteria and fungi compared to monocultures. In *T. grandis*, for example, the F layer harboured 3–4-fold higher populations of bacteria than in L layer, while the humus layer had lesser population density than in the F layers. On the other hand, an increasing trend was observed towards the humus layer in *E. globulus* at elevations of 500 m. similar to what was seen in polyculture litter. Fungal population density was highest in humus in all instances, except in the humus of *T. grandis* litter at elevations of 175 m. These observations suggest that microbial populations tend to vary in abundance in litter from diverse sources and can also vary with the altitudes. Gradual build-up from initially low inoculum sizes of maximum numbers in the fermentation layers, seem to be a general pattern followed by either an increase or reduction at the humus layer to numbers, however, more than what is recorded in fresh fallen litter. In more resistant litter as determined by the biochemical constituents of litter, the development is slower and the number may vary over a period of time (Rai and Srivastava 1982). This initial rapid increase in bacteria and fungi is due to availability of large amounts of nutrients in fresh litter. Microbial breakdown of litter releases carbon skeletons for further uptake and these include glutamate which can be readily utilised. Diversion of nutrients of fungal growth enhancement by the initial inoculum (Ling-Lee *et al.* 1977) on the other hand, leads to immobilisation of energy sources that cause brief periods of retarded population increases of the other saprotrophs.

One of the reasons attributed to low rates of *Pinus* litter decomposition is the presence of ectomycorrhizae that effectively divert nutrients from leachates and even reduce the quality of leaching which subsequently maintains low rates of microbial action (Gadgil and Gadgil 1975 and 1978). Two important facets of nutrient cycling include: (i) durations of maximum uptake and (ii) their residence time within the litter. Spatio-temporal fluctuations in litter biomass accumulation are a function of the phenological changes related to energy allocation for leaf production and shedding cycles with an important bearing on the contribution of nutrients to the understory. For instance, in the needles of *Pinus radiata*, N_2, P and K concentrations decrease with age, while Ca levels are maintained high. Decomposition of such needles contributes to accumulation of Ca (Forest and Ovington 1970; Reddy and Venkataiah 1991). Litter and foliar nutrition are also linked to the soil type that functions both as the source and the sink in the flow of nutrients (Turner and Lambert 1986). Sclerophyllous leaves bear significantly lesser levels of nutrients and undergo very slow rates of decomposition.

Litter Decomposition and Faunal Interaction

The accumulation of litter at the soil surface results in the colonisation of a variety of organisms, whose feeding activity and metabolic conversions result in further degradation of organic matter, so that the end-product of physical and chemical transformation is the humus, the production of which is largely mediated by soil microflora and fauna.

Soil fauna as such play a regulatory role in mineral cycling in the decomposition process, due to their effect on soil microbial activity either by direct consumption of microorganisms or changing structure through litter comminution. The outcome of these interactions between microbivorous soil arthropods and microorganisms in decomposition and mineral cycling at the microsite level shows a spatial and temporal fluctuation which can cumulatively effect fluxes at the ecosystem level. Temporal changes may be due to density variation in the consumers followed by a change in numbers and in the activity of fungi and bacteria (Faber and Verhoef 1991). These animal density variations often relate to seasonal cycles in the numerical size of microbivorous populations. It has been postulated that the soil arthropods with similar feeding, as a result of the microstratification in the soil, affect decomposition and mineralization processes differently (van Straalen *et al.* 1985; Faber and Verhoef 1991). With the microbes as the pioneers, microarthropods mostly insects and mites, symphylids and pauropods assist in the early stages of decomposition. Earthworms and large saprophages increase the rapidity of the breakdown. Detritus feeding soil animals follow-up decomposition initiated by microbes. The interactions between the flora and fauna increase the amount of nitrification. The organic matter ingested by the faunal elements is voided as faecal pellets which support rich microbial growth that again assist in the conversion of organic matter (Verhoef and Meinster 1991). Comminution, channelling and mixing, excretion and grazing are the essential modes of activity of the microarthropods that mediate physical and chemical breakdown (Gossley 1977). Litter is broken to nearly 100–200 μm sizes by the feeding activity of mesoarthropods and to even lesser dimensions of 20–50 μm by the microarthropods (Swift *et al.* 1979).

Grazing by microarthropods on fungi results only in selective removal of microbial populations without unduly destroying the resource. This selective removal triggers rapid increases in microbial populations which subsequently aid in the maintenance of the grazer population density. Such cyclic oscillations are, however, strongly determined by the nutrient status of the litter (Durham 1989). Microarthropods such as Collembola, Acarina, thrips, symphylids and pauropods are abundant in different layers of litter and their abundance is influenced by several factors including the level of moisture, nutrients and the prevalence of secondary plant substances in the chemical environment. That the soil microarthropods cause fragmentation of litter and subsequently mediate nutrient cycling is evident in several instances (Swift *et al.* 1979; Seastedt 1984). Litter bag experiments have adequately proved that the abundance of microarthropods plays a significant role in the seasonal dynamics of nutrients during the process of decomposition.

Nutrient outputs are directly increased by the consumption of litter (Reddy and Venkataiah 1989; Crossley 1977; Crossley and Hoglund 1962; Seastedt and Crossley 1980) and indirectly enhanced by fragmentation that increases the surface area by several folds for enhanced microbial colonisation and accelerated mineralisation. Correlations between microarthropod abundance and mineral concentrations have been documented with reference to the situation prevalent in *Eucalyptus* litter (Reddy and Venkataiah 1989) wherein rapid changes in the biomass of litter could be directly correlated to the abundance of Collembola and mites. These two groups which gain preponderance over the other groups of entomofauna in most instances constituted more than 86% of the insect population, and in terms of the fine-mesh and coarse-mesh bags used, Collembola represented 53.7% and 56.8% respectively, and Acarina accounted for 32.7% and 30% respectively, being more in the coarse mesh bags. Increased mobilisation and release of Ca and N_2 and possible uptake of P and K are evident in direct proportion to the activity of microarthropods.

Available information clearly indicates that the feeding activity of termites (Lee 1991; Buxton 1981) and other mesofauna result in weight loss from coarse-mesh bags. Termites in large numbers feed on cellulosic substances and through symbiotic digestion are able to utilise N_2-rich fractions from litter, very effectively. They consume 17–90% woody litter and up to 75% of total grass and forb litter. Foraging termites increase spatial heterogeneity of organic matter by their active movements and in the process of releasing their wastes. Similar functions are also attributed to the role played by ants that comminute litter (Reddy 1984). In areas where trees were planted in a once grassland area, vertical distribution of microarthropods showed that about 76% of the total microarthropods were recorded from surface litter layers and 24% from the fermentation and humus layers.

Correlational analysis point to significant relationships between the levels of primary metabolites and insect populations in litter suggesting that such changes are brought about by the feeding activities of the entomofauna. An initial population size of the insects was to increase significantly in the F layer concomitant with a reduction in the levels of utilisable metabolites. This was observed particularly in the levels of carbohydrates, proteins, amino acids, water content and N_2 which were high in fresh litter. With progress in decomposition, a gradual reduction in the levels of these nutrients was observed. While carbohydrates were lesser by 19–70%, proteins tend to be in equal levels in the F and H layers, subsequent to a reduction by 50%. Deletion of N_2 from litter was also obvious. Differences in the levels of amino acids tend to range only from 6 to 22%, suggesting their gradual accumulation in

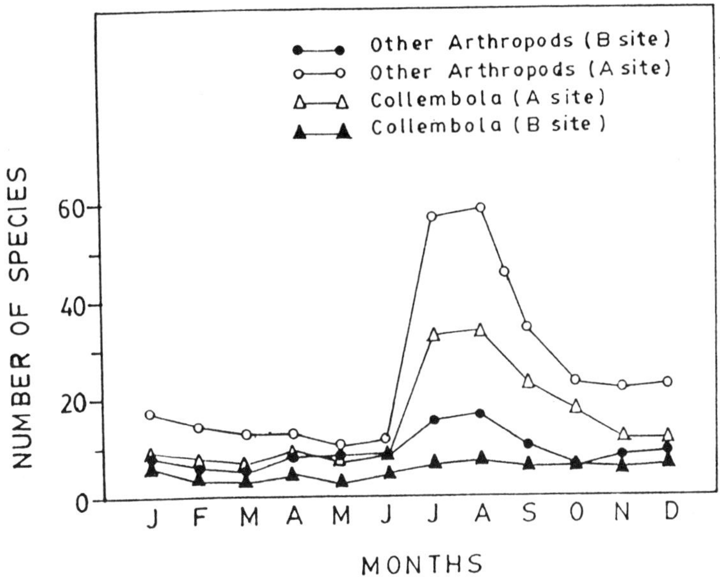

Fig. 6 Fluctuation in the abundance of Collembola and other arthropods of two different sites.

the H layer. Total phenols increased with decomposition by 27–28%. Of the minerals, Na and Mg exhibited a decreasing trend with decomposition, while Ca and K were higher in the F layers.

Recent investigations on the relative abundance of insect fauna in the litter of *Tectona grandis* (Ananthakrishnan *et al.* 1993) have documented site-specific variations which could be correlated with the changes in the biochemical profiles of the fresh litter, partly decomposed and fully decomposed humus layers. Significant differences were also recorded in relation to altitude and the soil type over which the site occurred. Seven groups of insects were identified in association with the litter of teak at mean see level (MSL) (Fig. 6). Mites dominated the assemblage in fresh litter followed by the collembolans, the scarabaeids, earwigs and staphylinids, termites and pseudoscorpions. The abundance of all the insect groups increased in the F layers. Maximal increases were recorded in the case of collembolans (8–9-fold) which dominated the assemblage, followed by the other groups (2–3-fold). The termites were abundant in the humus layer, and the staphylinids also exhibited an increasing trend in abundance, whereas all the other five groups were recorded in distinctly lesser numbers than those in the L and F layers.

The pattern of changes in the levels of metabolites was similar to what was observed as MSL. However, the extent of depletion was lesser towards the onset of decomposition of the F layer, followed by further reduction in the levels in humus. The initial levels of metabolites in litter were higher at this location that what was recorded at MSL. The residual nutrients at the H layer too were higher, except of amino acids, the depletion of which was observed to a greater extent. The level of N_2 was higher at the F and H layers compared to fresh litter, in contrast to a gradual loss of N_2 with decomposition. This points to the heterogenous nature of shifts in N_2 balance and N_2 to H_2O ratio in litter. The levels of Ca and Mg were significantly higher in the F layers. With the reduction of Ca through decomposition, Mg tended to increase 2–3-folds in the H layers. No difference was noticeable in the level of Na which, however, was much less in concentration (68 mg%) compared to the situation in litter at MSL. An upsurge in the level of K was recorded in the F layer followed by a reduction in the humus layer. It is accordingly evident that the mineral constitution varied and the levels of primary metabolites were marginally higher at higher altitudes. The dominant groups of insect fauna varied, though fluctuations in abundance followed comparable patterns in respect of decomposition.

Collembolans and mites are the dominant insect groups in litter (Ananthakrishnan 1979) playing important roles in decomposition and mineralization processes. Succession in collembolan communities is related to temporal availability of nutrient resources including fungi, and they increase in numbers when moisture and organic matter are available in plenty. They abound in the litter of moist rain forests with their main area of trophic activity confined to this layer wherein Collembola contribute significantly to changes in C/N ratios. This has been amply demonstrated in experimental micro-cosmos wherein the mobilisation of carbon was up to 28% in litter associated with *Gyptophygus thermophila*, compared to only 20% in their absence (Guru and Panda 1991). Enhanced respiratory rates and dehydrogenase activity in metabolic conversions in litter have also been reported (Teuben and Roelfsma 1991). *Tectrodontophora bielanesis* causes lipid degradation in both fresh and partly decomposed litter, playing important role in mineralisation (Durham 1989). The release of nitrates and ammonium is enhanced 3–4-folds when litter is associated with actively feeding Collembola (Lee 1991). Mesofaunal aggregations in decomposing spruce (*Picea abies*) are normally dominated by mites (150 per bag) compared to only 20–30 Collembola per bag. While initial colonisation is by relatively larger number of Collembola, subsequent increases are mainly of the mites including prostigmatids and astigmatids. Successional changes indicate initial colonisation by surface dwellers, followed by a preponderance of such insects that freshly migrate between the layers of litter tending towards advanced state of decomposition characterised by rapid uptake of C, followed by N_2 from the substrate nutrient pool.

Interestingly, mites dominated the assemblage of insects in fresh litter at altitudes of 150 m similar to the situation prevalent at MSL. Sharp increases in the abundance of Collembola with transition in the F layer followed similar patterns as at 250 m and MSL, suggesting the propensity of Collembola to easily attain large numbers with greater mobilisation of nutrients. The staphylinids which were absent at 250 m were observed at 1,350 m, which along with Dermaptera exhibited a reducing trend in abundance towards the H layer. The scarabaeids unlike in the other two situations, also were in large numbers in humus. Increase in termite abundance was similar to the pattern observed at MSL. While carbohydrates, proteins and amino acids were higher in fresh litter at 1,350 m altitude (by 12%, 33% and 32% respectively), subsequent changes with decomposition followed similar patterns of reduction. The water content was marginally less and total phenols tended to aggregate in higher quantities in humus, than at 250 m. While Na and K tended to increase with progress in decomposition, Ca and Mg were detected in lesser quantities, pointing to differences in patterns of release from litter. Of all the insect groups, the termites exhibited a capacity to not only tolerate the relatively high levels of total phenols, but have also successfully increased in numbers. While the collembolans and mites exhibited comparable patterns of rise and fall in abundance at all the three sites, the termites and Dermaptera were in relatively less numbers in the humus at 1,350 m. The latter two groups, however, showed an increase in numbers with progress in decomposition at MSL and 250 m. The staphylinids and scarabaeids fluctuated in contrasting patterns at MSL and at higher altitudes.

The role of litter inhabiting arthropods in comminution and physical breakdown of litter cannot be overemphasized (Butcher *et al.* 1971; Crossley 1977; Seastedt 1984) and the impact of environmental factors in determining their abundance and dynamics is well documented (Anderson *et al.* 1983; Reddy 1984; Wallwork 1959; Witkamp and Drift 1961; Seastedt and Crossley 1980; Swift *et al.* 1979). However, correlations of the dynamics of arthropods with the levels of available primary nutrients, essentially derived from litter, will provide useful indicators of the biochemical changes in decomposition, also in relation to breakdown products including the phenolic acids. Elevated levels of phenols deter herbivory in the context of which it is important to recognise the relationship between increasing levels of phenols in respect of litter decomposition and their accumulation in terms of the abundance and activity of the insect community found in association with litter. Such an accumulation may shift the balance away from continued feeding on litter. Observations have highlighted the significance of elevated levels of phenols in the humus layers, with a direct inverse correlation with insect abundance. This coincides with reduction in the levels of primary nutrients and shifts in H_2O/N_2 ratios. The accumulation of phenols tend to be higher in humus associated with clay-loam soil at MSL, compared to a reduction in the humus at 250 m and 1,350 m altitudes. Lesser accumulation could be a result of also the slower rates of decomposition at higher altitudes attributed to lesser environmental temperatures. The abundance of Collembola and mites as the dominant faunal elements in the L and F layers, followed similar trends of fluctuations as in the case of other groups. Succession in collembolan communities is also related to the temporal availability of nutrients. This contributes significantly to changes in N_2 ratios (Lee 1991) and promotes mineralisation. The active role of Collembola in decomposition processes was evident from the declining values of C/N ratio which is an important parameter of the rate of decomposition. Guru and Panda (1991) showed that the rate of decomposition was high in litter samples and low in soil samples where organic matter content was the lowest. The decrease in C/N ratio was mainly attributed to the loss of carbon.

3

Nutrient Cycling Patterns in Some Tropical Forest Litter

Forest systems irrespective of altitudes always harbour a larger diversity of Collembola and mites (Mukerji and Singh 1970; Ananthakrishnan *et al.* 1993). Mesofaunal assemblages in litter of deciduous forests of north-eastern India are dominated by mites, which can also tolerate wide amplitudes in fluctuations of environmental variables. Man-made forests such as those of *Shorea robusta* in the deciduous realms invest more energy in net primary productivity. The efficiency of energy storage and conversions monoculture stand paradoxically tends to be almost twice that of a natural forest (Ramam 1975) (Table 3). Tropical deciduous forests are not as complex as tropical rain forests in vertical organisation and species richness. However, litter accumulation touches peak values in April–May and the rate of litter disappearance is faster in warm and wet seasons. Of particular interest is the fact that the forests store considerable amount of essential nutrients (Misra 1972) (Ca—13, 310 kg/ha, N_2—9378 kg/ha, P—551 kg/ha) within half-a-metre soil depth. Due to the presence or inclusion of *Terminalia tomentosa* in the forest stand, soil fertility levels tend to become elevated.

Changes in the levels of minerals also varied significantly between the sites and did not follow any specific pattern. Ca was detected in marginally higher quantities in the F layers, similar to the trend registered in the decomposition of leguminous litter (Durham 1985). Subsequent reduction was registered in the H layers at all the sites irrespective of altitudes. The humus at 1,350 m was relatively rich in Na, while Mg tended to accumulate in the H layers at MSL. K being very mobile is less bound to nutrients and hence microarthropods have little impact on K depletion (Coleman *et al.* 1983).

Yet another interesting example is the dynamics of nutrient flow from the litter of *Eucalyptus*. Nutrient return through litterfall and stemflow tends to be in considerable proportions enhancing humification and rapid incorporation into the soil matrix. Litter production averages

Table 3. Energy stored in live and dead organic matter in natural- and man-made forest stands (after Ramam 1975)

Energy stored in litter	Natural deciduous forest	Man-made forest
Leaf litter	2.205	2.90
Non-leaf litter	2.08	—
Below ground biomass	14.784	30.08
Above ground biomass leaf	81.735	119.36
Net 1° productivity (K cal/m^2/yr)	4015	7665

3,500–6,000 kg/ha releasing 30–59 kg/ha of N_2, 2–4 kg/ha of P, Ca—15–51 kg/ha of Ca and 5–9 kg/ha of Mg. The quantity of Ca in particular is generally higher than what is comparatively released from temperate hardwood and conifers; the rate of release is almost twice as fast as in teak and is far more than 4.4 kg of Ca/ha released in the litter of *Shorea robusta* (Sharma 1971). Interspecific differences are, however, pronounced in terms of nutrients release and particularly N_2 and P contents tend to vary significantly. K inflow as an index of nutrient enrichment is normally in the range of 0.38–0.5 mg/100 g soil. Recent investigations on the biochemical changes in litter breakdown as identified in the L, F and H layers of *Eucalyptus globulus* have indicated significant differences in litter biochemical contribution in a layer-specific mode (Ananthakrishnan *et al.* 1993). Litter systems tend to be complex both in terms of substrate heterogeneity and the nature of decomposed assemblages, which are vastly influenced by the biochemical characteristics and physical conditions of the milieu. Collembola and fungi compete for resources, further contributing to the complexity in the dynamics of both microflora and microfauna, indirectly by choosing to exploit a common nutritional source. While collembolan grazing promotes an adaptive increase in the dispersal of fungi through phoresis on physical contact, fungal colonization contributes to metabolic changes in the substrate (Fox and Macauley 1977).

Litterfall and concomitant return of nutrients to soil contribute to abiotic buffering in soils. Monoculture of *Tectona grandis* effectively deposits 1.7–17.2 kg/ha/N_2/m with proportionately high releases of phosphorus (0.2–3 kg/ha/m) and potassium (0.7–10.2 kg/ha/m) (Srivastava, Kaul and Mathur 1972). Annual return to the soil tends to be 10–20-fold higher when the plant cover is a complex forest or even in the case of polycultures including *Terminalia* and *Anogeissus latifolia* along with teak (Seth, Kaul and Gupta 1963). Changes in C/N ratios significantly alter the rates of decomposition. For instance, the characteristically high levels of ash (15–20%) with low C/N ratio and lignin content in *Tectona grandis* and *Terminalia tomentosa* contribute to lesser rates of decomposition. Lignin contents of 30% dry weight significantly contributes to very retarded rates of decomposition as evidenced in the case of *Buchanania lanza* and *Shorea robusta* (Singh 1969). Initial phases of decomposition are characterised by rapid losses of less complex carbon compounds such as sugars, hemicellulases, pectins and organic acids, whereas N_2-rich compounds tend to increase in relative abundance in litter reducing the C/N ratio with progress in time. In oak, for instance, a initial C/N ratio of 71 reduced to 28 in an year (Gilbert and Brock 1968). In soils the ratio is generally 10–12 wherein mineralisation is almost complete. Litter decomposition spans 2–7 months in temperate deciduous forests (Ovington 1962). Other extremes include 28–60 years in black oak and ponderose pine (Jenny *et al.* 1949). Gupta and Rajvanshi (1991) indicate that the decomposing litter of *Dalbergia sisoo* in dry deciduous forests of North India are characterised by high N_2 (1.2–2.6%), and water soluble carbohydrates (16–20%), with C/N ratios of 18–15. The rate of decomposition varies from 1.6 to 2.7 mg/day and in tropical conditions the rate of decay is from 1.07 to 13.1 mg/day (Singh and Gupta 1977).

Nutritional Cycling Patterns in Some Tropical Forest Litter

Biologically mediated immobilisation and mineralisation contribute significantly to the dynamics of nutrients and minerals in soils (Cole *et al.* 1978; Van Veen and Frissel 1981). The low molecular weight amino acids, mono- and oligosaccharides which are elements of the fast decomposition cycles, include polygeneric nitrogenous compounds including chitin, organic phosphorus and cell wall components that are degraded at much lesser rates. Chitin on decomposition contributes to the N_2 pool as NH_4. Biochemical mineralisation is enzymatic, biased to specific needs of the organisms in the environment that readily degrade the organic matter contents constituted by polyphenol-rich organics and carbohydrate-rich sources (Foster and Martin 1981). Resource utilisation and cyclic oscillations in population density of the nutrient absorbers, live-root feeders, decomposers and grazers followed by successional patterns towards a stable state in mature forests is a characteristic feature (Cole *et al.* 1978). Nurient cycling patterns in tropical forests are diverse and are influenced by such factors as biomass (Brown and Lugo 1982 and 1984). Cycling of N_2, P, K Ca and Mg are crucial as they are soil-derived nutrients and can also limit primary production and are themselves influenced by climate, species composition and successional status. The infertile soils, viz. oxisols and ultisols which are moderate to strongly acidic constitute the predominant soil types in the moist tropical forests.

The concentrations of N_2, P, K, Ca and Mg in foliage may vary with the type of soil ranging from 0.74 to 2.52%, 0.02 to 0.15%, 0.38 to 1.67%, 0.11 to 2.20% and 0.11 to 0.88% respectively (Medina 1984). Quantification of mineral mobilisation through litterfall serves as an index of organic matter conversions and preferential uptake of nutrients (Hirose 1975; Vitousek 1982). It is important to recognise that the nutrient status of litter is related to the fertility levels of the soils (Feller 1980). High nutrient concentrations are characterised by low organic matter to nutrient ratios. The levels of N_2, P and Ca tend to be significantly lesser in forest litter, over less fertile soils, indicating that immobilisation in the foliage in general is low in both N_2 and P. Decomposition in terms of litter turnover in relation to annual litterfall including the role of invertebrate and microbes, is slower in montane forests due to lower temperatures (Tanner 1981), and is comparable to rates of decay over infertile soils. This is in contrast to the high rates in lowland moist tropical forest litter. The mobilisation of N_2 as NH^+_4 and NO_3 is known to be faster in tropical soils than in temperate soils. Low N_2 and P as major factors in tropical conditions and high tannins particularly in temperate situations are implicated in relation to the observed lesser rates of decomposition (Melilo *et al.* 1982).

The relative levels of the important minerals in the leaves of trees that dominate the assemblage in tropical deciduous forests, of which highest levels of P are found in the foliage of *Terminalia tomentosa* and N_2 levels are least in *Anogeissus latifolia* are indicated in Table 4 (Ramam, 1975; Deshbandu 1975). Even as nutrient concentrations in litter have a vital role in determining their subsequent uptake by the forest stand, they also determine the trophic dynamics of the animal consumers specially of the comminutors represented mainly by the insect fauna. The impact of the relative abundance of primary and secondary nutrients prevalent in the litter significantly contributes to either rapid uptake by the consumers or even deters the grazers (Anderson *et al.* 1983; Ananthakrishnan *et al.* 1993).

No specific relationships are discernible between nitrogen and phosphorus in the forest floor. Tropical forests have lesser amounts of phosphorus and nitrogen levels directly proportional to the biomass on the forest floor. The demand from the mircrobial component for N and P is rather constant and maintains a ratio of 10 : 1 for steady progress of decomposition (Alexander 1977). N and P concentration in litter and microbial biomass tend to be high and when organic matter complexes with nutrients, their mobilisation is retarded. P circulation tends to be higher in tropics which is directly proportional to the quantity of litterfall.

Table 4. Foliar analysis of some tropical deciduous trees (%) (after Raman, 1975)

Trees	Ash	Ca	Mg	P	N_2
Shorea robusta	6.9	1.65	1.97	0.42	2.0
Anogeissus latifolia	8.9	2.82	2.01	0.58	1.90
Terminalia tomentosa	9.9	2.56	0.83	0.56	2.4
Buchnania lazann	13.0	1.5	–	0.17	2.1
Diospyros tomentosum	5.95	2.01	–	0.32	2.4
Storage of Ca and N_2 (kg/ha) in litter					
		Natural forest		*Shorea* plantations	
		Ca	N_2	Ca	N_2
Leaf litter		99	111	66	74
No-leaf litter		85	16		

The mean residence time of litter nutrients ranges from 0.9 to 1.6 years in tropical broad-leaved deciduous forests as against 2.7 to 4.8 years for organic matter and N_2 in warm temperature forests. In colder climates the residence time is as high as from 10.2 to 19 years and extremes of 32.8 years are recorded for needle decomposition (Vogt *et al.* 1986). Soils with low H_2O-holding capacity have lesser N_2 and P than what is found in the litter (Lewis and Harding 1963) related to poor retention capacity of soils. Conservation of nutrients and efficient utilisation are related to the dynamics of the biotic components which constitutes the most important mechanism of mobilisation in evergreen forests where the forest floor is the major site of nutrient retention. In deciduous forests, above ground biomass is the dominant conservation site. The organic matter content is approximately 11% (dry weight) in montane evergreen forests, 7.3% in semi-deciduous teak forests and 5.7% in deplerocarp-dominated stands, specially confined to the upper 15 cm of soil horizon. There is an increase in C/N ratio both in the upper and lower soil horizons in relation to altitudes over 500 m. While a primary forest is known to accumulate 10,000 kg/ha/yr of N_2 in the humus, exploited forests retain only about 400 kg/ha/a. Several confirmations are available to show that 8,000–9,000 kg/ha/a N_2 is normally available in the upper 90 cm of soil (Cole *et al.* 1987). Significant increases in the levels of N_2 in litter result from contributions from aboveground sources including through fall, stemflow and inclusions from faunal sources as frass. Green leaf material contains water soluble fractions that contribute to litter N_2. With N_2 levels ranging from 0.3 to 5% of the dry matter during degradation of litter, subsequent increases are also related to the ionic balance contribution to permeability levels of microbial cell masses and their grazers' metabolic requirements. It is in this context that soil fauna play a crucial role in cropping microbial populations, indirectly regulating litter decomposition rates. The end products of litter breakdown including phenolics determine the biochemical suitability of the nutrient source (Hansley 1954; Gilbert and Bocock 1960; Usher 1985; Ananthakrishnan *et al.* 1983). With progressive demonstration, the accumulation of recalcitrant organic acids, combines with the relatively low calorific value of degraded litter and thus regulates the feeding dynamics of the faunal elements.

Incidentally mention may be made of the fact that organic and inorganic substances are leached from the tree canopies, the organics being in the form of soluble carbohydrates. Potassium is leached considerably in tropical forests and different tree species differ in the degree of leaching and the loss by leaching depends on the age of the leaves. A greater loss of nutrients occurs from older than younger leaves and hence an understanding of the chemical constituents—both organic and inorganic is essential. Besides fresh fallen and degrading litter, it would be useful to also have the chemical components of maturing and senescing leaves examined, so as to make a comparative assessment of the nutrients and their transfer. It is important to also recognise that nutrient withdrawal is progressive, even prior to abscission, particularly during senescence. This is specially so of Ca and P, the translocation of which is not reversible. Hence Ca and P levels as ratios of total nutrient status serve as useful indices of selective uptake and subsequent release into the nutrient pool in the process of litter decomposition. The continuous addition of decomposing organic matter and inorganic nutrients in the layers of litter has to be evaluated quantitatively also in relation to microbial activity. Interestingly enough, there is evidence that suggests a correlation between rates of litter decomposition and the base content. Most potential decomposers are inactive or less active at pH below 5. The base content of litter varies through the year (Lathwell and Grove 1986). In forest floors with heavy undergrowth, the C : N ratios are recorded at 40 : 1, indicating a carbon bias. Release of allelochemicals in proportionately higher quantities, and of complex energy demanding substances including recalcitrant phenolics, further inhibit N_2 fixation (Robertson and Vitousek 1981). Nitrification in mixed deciduous forests is directly proportional to the availability of NH^+_4 released by the heterotrophs that gain preponderance over other trophic elements for most periods of time. This situation corresponds to periods of high C : N ratios. A direct correlation has been indicated with high moisture levels contributing to the maintenance of favourable temperatures in the litter matrices. This leads to greater mobilisation of organic matter derived from diverse litter sources, which subsequently enhance nitrification and abiotic buffering (Doran 1980) crucial

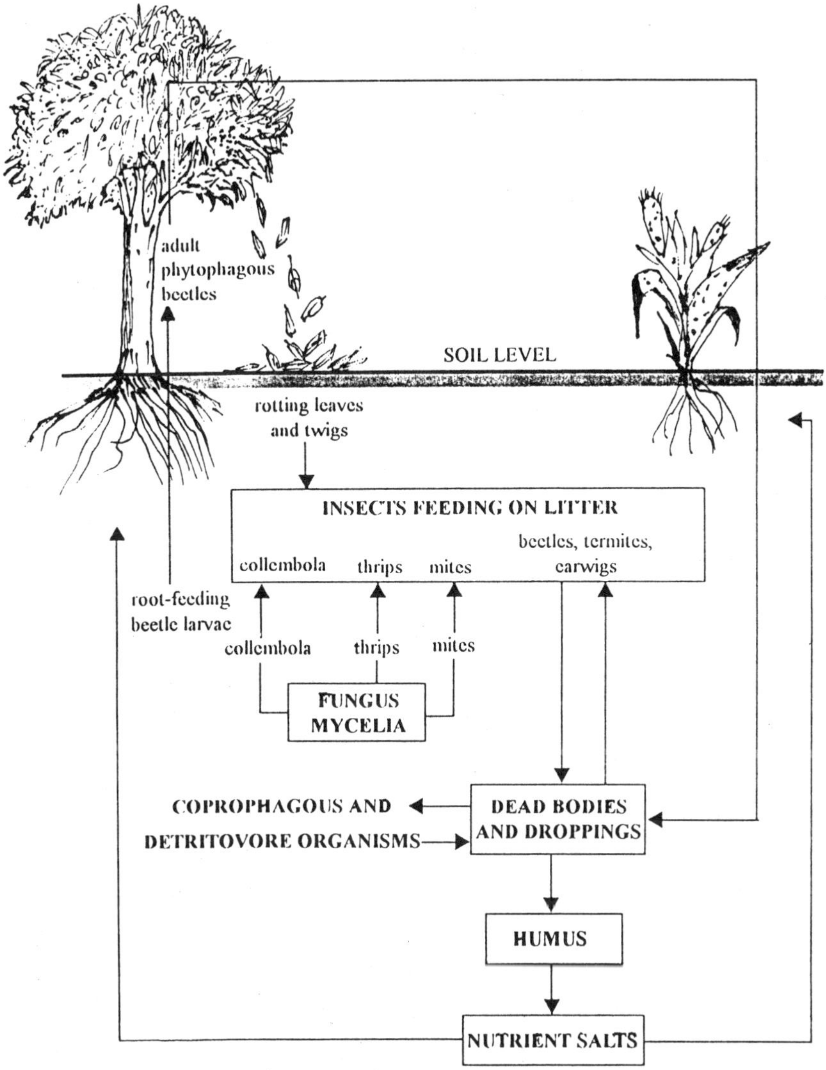

Fig. 7 Schematic diagram showing some food relationships of soil organisms and the part they play in the process of humification.

for the persistent operation of the diverse nutrient cycle pathways. Tropical forests that are very productive are rich in nutrients in terms of diversity and accordingly the patterns of nutrient cycling tend to be diverse. Nutrients most likely to limit primary production and other ecosystem functions are N, P, Ca and Mg. Soil fertility, species composition, climate and successional status are the major features of importance in governing variations in nutrient cycles. Total nutrient contents are, in turn, determined by biomass and their distribution in different plant parts. Higher contents of water soluble substances promote decomposition, while high lignin content makes it difficult. Greatest influence is perhaps exercised by C/N ratios presumably because C and N serve as nutrients for organisms participating in decomposition. By promoting recycle of nutrients, the microflora, and entomofauna play an effective role in the preservation and productivity of forests (Fig. 7).

4

Microbial–Microarthropod Interactions in Litter

Litter catabolism in soil depends on microorganisms, the soil faunal elements tending to act as catalysts enhancing energy and nutrient influxes. Needless to emphasise that these feeding activities increase the surface area of the substrate exposed to microbial attack. Quantitative changes in the soil microflora are brought about by the grazing effects of the litter faunal elements maintaining fungal and bacterial population in active growth stages. Among the litter arthropods, Collembola are one of the most abundant groups, followed by the mites and thrips feeding mainly on the detritus, fungi and bacteria. *Folsomia candida* is an interesting example whose symbiotic interaction with bacteria enhances the use of chitin. Since this species feeds on its own chitinous exuvia, it has an efficient mechanism for utilisation of chitin (Borkott and Insam 1990). Compared to the numbers of chitin-digesting microorganisms in litter/soil, the numbers found in the gut and faeces of *F. candida* is very high.

Microarthropods are known to control the distribution, abundance and activity of litter fungi and bacteria, through selective grazing of fungi and their ability to disperse fungal inoculum. Selective grazing while inducing competitive interaction among grazers, serve to disperse propagules of the grazed species and their excretion may control the rate and proportion of nutrient return to the grazed site (Lussenhop 1992). Hence, essentially microarthropods, in particular the Collembola, Protura, and mites function as grazers and transporters. Selective grazing may have marked effects on ecosystem nutrients cycling and may result in a population shift of fungi. Further, feeding by soil mesofauna leads to a shift from fungal to bacterial biomass, since mechanical disruption and mixing with other soil micronutrients may be more beneficial to bacteria than fungi (Anderson and Bignell 1980). For transport, even though mycangia may be lacking, many Oribatid mites have sculptured integument which could function as mycangia. The high surface volume of hyphae and hyphal network enable translocation of nutrients to sites of active decomposition. Larval numbers tend to increase with the increase in fungal species. Since microarthropod habitats such as litter layers are rich in fungi and bacteria, products of microbial degradation and exoenzymes are equally rich, enabling digestive adaptation, with the result that these gut microorganisms enable Collembola to produce more eggs when fed with fungi. Cellulases commonly occur in the guts of fungal-fed microarthropods, while bacteria are associated with chitinases, proteases and amylases (Stefaniak and Seniezak 1981). Microarthropods may facilitate the activity of fungal enzymes in their guts. They also return considerable amount of mineral nutrients in faeces.

Microbial activity is stimulated by selective grazing of fungi and dispersal of fungal inoculum by litter/soil microarthropods results. Selective grazing can control species distribution and favour decomposer fungi (Fig. 8): Moderate grazing may stimulate microbial grazing augmenting either mineralisation or immobilisation of nutrients by microorganisms. True edaphic genera such as *Onychiurus* are predominantly

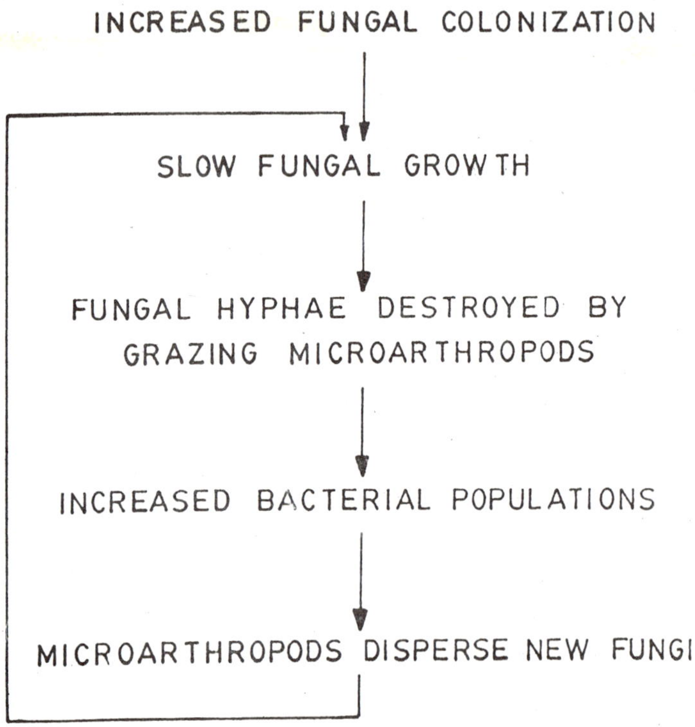

Fig. 8 Selective grazing and microarthropod dispersion of fungi.

mycophagous, while hemiedaphic genera such as *Tomoceros* and *Isotoma* contain large quantities of fungal diet in their guts. Mycophagous tubuliferous thrips abundant in litter also harbour large quantities of fungal species in their gut such as *Anthostomella, Pestalotia, Melanographium, Lasiodiplodia*, to mention a few (Ananthakrishnan and Suresh, 1983).

Substrate quality is the major factor which determines the rate of leaf litter decomposition, the diversity of microorganisms associated with litter, composition of mycoflora as well as that microarthropods such as Collembola, mites and thrips. The primary colonizers of litter are Coelomomyces and most of the initial colonizers, often dominant, are *Aspergillus* and *Penicillia* in teak and eucalyptus litters. The initial colonizers also include *Curvularia, Cladosporium* and *Alternaria*. The secondary saprophytes of litter comprised several genera of fungi imperfecti and a few genera of Ascomycetes and Basidiomycetes. During advanced stages of decomposition Zygomycetes are also involved, indicating a general scheme of fungal succession. Freshly fallen litter has species of *Penicillium*. The abundance of *Aspergillus* and *Penicillia* may be attributed to their ability to grow over a wide range of temperatures, pH, moisture content and a range of substrates. Of the fungi isolated from teak and eucalyptus litter, 79–87% are fungi imperfecti, 9.4–12% Zygomycetes. *Aspergillus, Penicillia, Tritirachium, Coniella, Myrothecium, Chaetonella* species are dominant primary colonizers. Among the secondary saprophytic-fungi are *Aspergillus flavus, A. niger, Curvularia lunata, Mucor* sp., *Fusarium, Alternaria* and *Myrothecium. Trichoderma viride* was frequent in the later stages of decay (Mary and Sankaran, 1991).

In studies on a heterogenous forest litter ecosystem as represented by Idikki and Munnar forests of Kerala, species richness and abundance of the fungal species were represented by 11–13 species at both sites (Tables 5, 6). *Aspergillus, Mucor, Rhizopus* and *Fusarium* were the dominant species in all the layers. Patterns of species abundance varied considerably in the L layer (12 species at Idikki and 7 species at Munnar) and more or less similar to the F and H layers at both sites. This may be attributed to the levels of metabolites, minerals, moisture and successional changes in respect of litter breakdown. Species of *Gliocladium, Curvularia* and *Absidia,* were specific to Munnar and of *Emericella, Chrysosporium, Beltraniella, Cylindrocladium* and *Aureobasidium* to Idikki litter indicating the site-specific species diversity of fungi (Fig. 9). Drastic decline in the abundance and heterogeneity of fungal community was common to the humus layer irrespective of altitudinal variations emphasizing the significance of

Table 5. Fungal species and number of colonies in association with L, F and H layers of natural forest litter at Idikki and Munnar forests, Western Ghats.

L-Layer		F-Layer		H-Layer	
Colonies	Number/g	Colonies	Number/g	Colonies	Number/g
Idikki					
Aspergillus niger	23 ± 1.2	*Aspergillus*	21 ± 0.2	*Aspergillus*	14 ± 1.1
A. nidulus	10 ± 0.8	*Mucor*	18 ± 1.7	*Fusarium*	13 ± 0.9
Mucor	24 ± 1.8	Rhizopus	08 ± 02	*Penicillium*	08 ± 0.6
Fusarium	20 ± 0.4	*Penicillium*	16 ± 1.7	*Mucor*	05 ± 1.7
Trichoderma	18 ± 1.2	*Fusarium*	16 ± 0.6	*Rhizopus*	05 ± 1.7
Pencillium	15 ± 0.7	*Beltraniella*	10 ± 1.3	*Aureobasidium*	03 ± 1.5
Rhizopus	10 ± 1.7	*Emericella*	0.3 ± 0.3	*Trichoderma*	02 ± 0.4
Beltraniella	05 ± 0.6	*Menoniella*	0.3 ± 0.8		
Cylindrocladium	0.3 ± 1.2				
Chaetomium	0.3 ± 1.1				
Chrysosporium	0.3 ± 1.2				
Munnar					
Aspergillus	08 ± 1.3	*Mucor*	15 ± 0.8	*Mucor*	15 ± 1.1
Fusarium	07 ± 0.8	*Aspergillus*	11 ± 0.9	*Aspergillus*	14 ± 0.8
Mucor	06 ± 0.4	*Fusarium*	11 ± 1.1	*Rhizopus*	08 ± 0.8
Penicillium	05 ± 1.3	*Rhizopus*	12 ± 1.7	*Fusarium*	05 ± 1.1
Absidia	05 ± 0.5	*Penicillium*	06 ± 0.8	*Alternaria*	05 ± 0.7
Trichoderma	02 ± 1.2	*Trichoderma*	05 ± 1.2	*Trichoderma*	04 ± 0.9
Gliocladium	02 ± 1.2	*Chaetomia*	04 ± 0.3		
Memnoniella	04 ± 0.2				

recorded primary metabolites and elevated phenolic levels (Ananthakrishnan and Sabu Thomas 1994).

This diversity of fungal elements in litter at varying altitudes is equally of interest, with the associated differences also in the faunal elements.

Interactions between Microbial and Faunal Elements

Interaction among fungi, bacteria and microarthropods are central to many processes in litter, from decomposition to the functioning of the rhizosphere. Grazing is a complex phenomenon involving consumption of parts of living organisms. Grazers may be selective and affect competition between grazed species, resulting in efficient dispersal of fungal propagules. Excretion by grazers may control the rate and proportion of nutrient return to the grazed sites. Collembola, Acarina, Protura form

Table 6. Fungal species and number of colonies in association with L, F and H layers of shola forest litter along varying elevations in the Kodaikanal Hills

Litter layer	1,440 m (Perumalmalai)		1,820 m (Shenbaganur)		2,220 m (Pambar)	
	Colonies	Number/g	Colonies	Number/g	Colonies	Number/g
L	*Rhizopus*	24 ± 2.1	*Aspergillus*	18 ± 0.2	*Rhizopus*	18 ± 0.2
	Aspergillus	22 ± 0.2	*Rhizopus*	17 ± 1.2	*Aspergillus*	16 ± 2.1
	Trichoderma	22 ± 0.8	*Chaetomium*	15 ± 0.8	*Penicillium*	16 ± 1.1
	Chaetomium	18 ± 1.1	*Trichoderma*	13 ± 1.1	*Trichoderma*	10 ± 1.8
	Emericella	11 ± 0.3	*Penicillium*	10 ± 0.3	*Fusarium*	8 ± 0.6
	Chrysosporium	10 ± 1.2	*Mucor*	8 ± 1.1	*Alternaria*	8 ± 3.1
	Mucor	9 ± 1.5	*Memnoniella*	6 ± 1.8	*Zygosporium*	6 ± 2.2
	Beltraniella	9 ± 1.0	*Gliocladium*	6 ± 0.7	*Phialocladium*	6 ± 0.8
	Penicillium	6 ± 2.2	*Beltraniella*	3 ± 1.8	*Aureobasidium*	5 ± 1.1
	Gliocladium	4 ± 1.2	*Cylindrocladium*	2 ± 1.1	*Emericella*	4 ± 2.3
	Cylindrocladium	4 ± 1.0	*Chrysosporium*	2 ± 1.8	*Absidia*	2 ± 2.1
	Memnoniella	2 ± 1.1			*Chaetomium*	1 ± 0.9
	Fusarium	2 ± 1.2				
F	*Rhizopus*	21 ± 3.1	*Rhizopus*	14 ± 2.2	*Aspergillus*	14 ± 2.1
	Chaetomium	17 ± 1.1	*Chaetomium*	13 ± 1.3	*Chaetomium*	13 ± 11
	Aspergillus	15 ± 1.8	*Aspergillus*	12 ± 1.8	*Trichoderma*	12 ± 1.8
	Fusarium	14 ± 1.2	*Penicillium*	10 ± 1.2	*Rhizopus*	10 ± 2.1
	Trichoderma	14 ± 0.8	*Beltraniella*	8 ± 2.1	*Fusarium*	9 ± 1.1
	Penicillium	12 ± 1.1	*Cylindrocladium*	6 ± 3.1	*Penicillium*	7 ± 2.1
	Humicola	10 ± 1.2	*Memnoniella*	2 ± 2.1	*Zygosporium*	5 ± 1.8
	Gliocladium	11 ± 2.1	*Mucor*	1 ± 1.1		
	Emericella	8 ± 3.1				
	Alternaria	6 ± 2.2				

Litter layer	1,440 m (Perumalmalai)		1,820 m (Shenbaganur)		2,220 m (Pambar)	
	Colonies	Number/g	Colonies	Number/g	Colonies	Number/g
	Mucor	5 ± 1.8				
	Memnoniella	4 ± 1.8				
	Aspergillus	12 ± 1.3	*Aspergillus*	8 ± 1.1	*Rhizopus*	6 ± 1.2
H	*Chaetomium*	10 ± 2.1	*Trichoderma*	5 ± 1.8	*Aspergillus*	6 ± 1.8
	Rhizopus	7 ± 1.4	*Rhizopus*	3 ± 2.1	*Chaetomium*	5 ± 1.2
	Trichoderma	5 ± 1.2	*Penicillium*	2 ± 1.1	*Penicillium*	3 ± 2.2
	Fusarium	5 ± 0.8				
	Penicillium	4 ± 1.1				
	Absidia	2 ± 2.1				
	Emericella	2 ± 0.7				

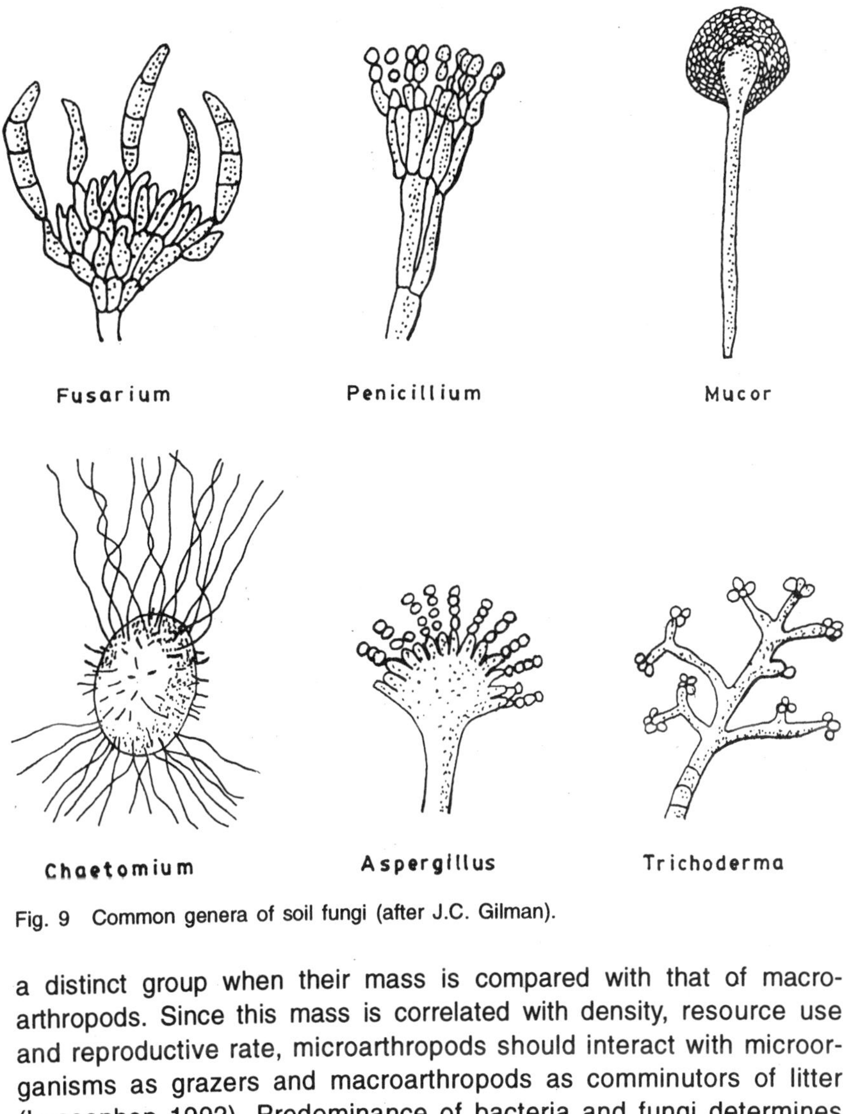

Fig. 9 Common genera of soil fungi (after J.C. Gilman).

a distinct group when their mass is compared with that of macroarthropods. Since this mass is correlated with density, resource use and reproductive rate, microarthropods should interact with microorganisms as grazers and macroarthropods as comminutors of litter (Lussenhop 1992). Predominance of bacteria and fungi determines invertebrate food web. While bacterial activity may be locally greater than fungi, hyphal growth is a central adaptation of fungi that determines their biology. The high surface and volume ratio of hyphae allows efficient utilisation of the products of external enzymes. Hyphal network enables translocation of nutrients to sites of active decomposition. It is therefore the vulnerability of hyphal network that is the basis for competitive interaction with invertebrates. As a result of these interactions there is an inverse relationship between fungal species number and decomposition.

Microarthropod grazing intensity is strong enough to control the abundance and distribution of fungi. Microarthropod activity favours growth of bacteria, probably by mixing cells with fresh substrates. Bacterial biomass exceeded fungal biomass when 10 or more Collembola were present, such as *Folsomia candida* (Choudhuri and Banerjee 1975). The stimulating effect of microarthropods on fungal and bacterial growth effect nutrient transformation. The litter inhabiting Collembola *Tomoceros minor* is known to be associated with nitrogen immobilization in litter.

Microarthropods are rich in microorganisms, microbial exoenzymes and products of microbial degradation. As a result, ingestion of a variety of microbially conditioned materials returns and triturition of food materials may be the most important digestive adaptation of microarthropods. Mouth parts of oribatids (*Phthiracaurus* sp.) and Collembola-like *Tomoceros longicornis* function to minimize the size of food particles, which tends to enhance the activity of microbial endogenous enzymes in the gut (Dinsdale 1974; Manton 1977).

Collembola can select hyphae with the highest nutrient content and produce more eggs when fed on fungi with high nitrogen content (Boolts and Anderson 1978) and a particular microbial species present in the gut effects maturity and number of eggs in oribatids (Haq and Konikara 1988). In nature, the contribution of microarthropods to the mineral pool is significant, since nitrogen concentration of faecal pellets was 56% higher than the hyphae (Verhoef *et al.* 1988). Microarthropods modify distribution pattern of bacterial and fungal species by dispersion of propagules from sites of sporulation. *Alternaria, Epicoccum* and *Penicillium* were more frequent in litter inhabited by microarthropods. The pattern of dispersion of propagules include: a) microarthropods carrying more propagules and species in litter, b) aggregation of spores at the sporulation sites dispersed rapidly by microarthropods which also bring fresh inoculum to the sites they have grazed (Lussanhop 1992).

Acarine–Fungal Relationship

Many groups of mites have evolved from an ancestral predatory life style into a variety of saprophagous, phytophagous and parasitic niches and have exploited diverse niches because of their small size. Several acarines have exploited fungi as food and developed symbiotic association between mites dependent on fungi (Table 7). Mycetobionts needing fungi for food, mycetophiles using fungi in many ways other

Table 7. Diversity in the mite fauna in natural and monoculture forest litter

Group	Percentage group composition			Family	Species	Trophic level
		Monoculture				
	Natural forest	Eucalyptus	Teak			
Cryptostigmata	43%	57%	55%	Plateremaedae	*Plateremaeus* sp.	Fungivorous
				Haplozetidae	*Paraxylobates imitans*	Fungivorous
					Zetorchestes saltator	Fungivorous
				Galumnidae	*Trachygalumna* sp.	Fungivorous
				Mesoplophoridae	*Apoplophora remota*	Fungivorous
				Pelopidae	*Eupelops*	Fungivorous

than food, and mycetoxenes occurring in fungi accidentally are ecological categories of acarine-fungal association. The importance of interaction between oribatid mites and fungi in the soil ecosystem has been often stressed in terms of nutrient cycling and physical decomposition of organic materials.

Several cryptostigmatid mites and mycophagous collembolans act as dissemination agents for fungal spores by passing through the gut or on the cuticle, across regions of litter that are impenetrable to the microbiota and hence acting as catalysts by bringing about primary decomposition. Behan and Hill (1978) reported that soil oribatid mites carried propagules from over 20 species of fungi and Hassall *et al.* (1986 and 1986a) found that *Onychiurus subtenuis* carried spores of over 200 species of fungi thereby increasing the decomposition rate.

Mycophagous Litter Thrips

Thrips are associated with fungi as mycetobionts being dependent on fungal hosts, or as facultative mycetophiles. The Uzelothripidae, Merothripidae as well as Idolothripidae are recognized fungal feeders, abundant in forest litter. The Phlaeothripidae also include mycelial feeders. In view of the abundance of many sporophagous species in more humid situations as obtained in the tropical rain forests, their tendency to aggregate in the larval/adult conditions or both, comprising larger number of individuals in a given unit area and their feeding pattern variability in terms of host preferences, it may be presumed that their potential to effectively consume large quantities of spores of plant pathogenic fungi is considerable. Such a concentrated feeding action by large number of individuals—adults and nymphs of many of these species could be effective in the natural control particularly of fresh fungus-attacked young plants. Further, the ability of many of these species to effectively reproduce by oviparity, ovoviviparity and viviparity and the high fecundity rates positively indicate an increasing possibility of such utilization. An equally important aspect is the dual role played by many of these thrips species, namely as mechanical vectors as well as efficient spore feeders of plant pathogenic fungi (Ananthakrishnan, 1981).

The build-up of such fungal inoculum on dry and dead logs/leaf litter and dead twigs at or near the sites of their congenial host plants was evident alongside with the respective sporophagous thrips species. *Dinothrips sumatrensis* and *Elaphrothrips denticollis* feed on the spores of *Lasiodiplodia theobromae* and *Pestalotia* sp., respectively, while *Elaphrothrips procer* was found to feed on both. *Bactrothrips idolomor-*

phus fed on spores of *Pestalotia* and an unidentified fungal species. Significant among these observations is the feeding by thrips on both the mature and immature spores by breaking open the pycnidia of these fungi, thereby exposing the spores for easier dispersal. Of equal significance is the observation on the preferential feeding and the amount of spores consumed by individuals of the thrips species. The reproductive potential of individuals within a given feeding site, in terms of the degree of aggregation of individuals is directly correlated with the spore consumption rate. In view of the large number of fungal spores consumed by an individual at a given time, the overall consumption of a large quantity of inoculum by a number of aggregating populations at a given area during the whole life-cycle is of considerable importance. This dual function of consumption and dissemination of pathogenic fungal spores appears an essential aspect of these sporophagous thrips.

In the warmer parts of the world many species of thrips inhabit leaf litter feeding on the associated fungi, either on the hyphae or on spores. Interestingly enough, all the phlaeothripines associated with litter are mycetophilous or hyphal feeders and the idolothripids essentially feed on fungal spores. The presence of more than 50 species of litter-infesting thrips in one small area of the southern Brazil forest appears exceptional (Mound 1977) at an altitude of 500 m, with an annual rainfall of 200 mm and temperature range of 0–38°C. The litter essentially comprises fallen leaves of various species, the dominant trees being members of the Lauraceae. This is an intermediate flora that has promoted the diversity of the leaf litter thrips, mostly confined to the upper layers in small localised concentrations. Mound (1977) has indicated that in Trinidad, eastern Australia and southern Brazil which are all tropical or subtropical with diverse flora and a marked variation in rainfall, the leaf litter harbours a diverse variety of thrips. Malaya, Java, eastern India including Assam, Meghalaya, the Western and Eastern Ghats, though tropical with the tropical deciduous and tropical humid forests, harbour comparatively fewer litter-infesting species in spite of the diversity of flora. Thrips are least common in wet litter or under trees with large leaves, but leaf litter produced by diverse flora with smaller leaves presumably has an open cavity structure with a large area of leaf surface exposed and bearing fungi on which thrips can move and feed (Mound 1977). Leaf litter provides a rather uniform habitat across large areas.

In Australia, leaf litter thrips occur in large numbers in dry litter such as those of eucalyptus in exposed sunny places. Very few species have been identified from the rain forest litter in Australia, although rain forest litter in Trinidad and Solomon islands (Mound 1972) contain

Table 8. Association of some sporophagous thrips with plant pathogenic spores in forest communities

Thrips species	Host	Fungus species fed	*Number of spores consumed			*Number of thrips/ agg-regation	Spore consumption potential of an aggregation of thrips
			IM	M	Total		
Bactrothrips idolomorphus	*Shorea robusta*	*Pestalotia* sp.	-	25	25	18	450
Dinothrips sumatrensis	*Mallotus albus*	*Lasiodiplodia theobromae*	203	162	365	12	4,380
	Hevea braziliensis		65	131	196	12	2,352
	Piper nigrum		1507	121	1178	12	14,136
	Mangifera indica		75	133	208	12	2,496
	Lannea coromandelica		77	341	418	12	5,016
	Anacardium occidentale		258	41	299	12	3,588
Elaphrothrips procer	*Tectona grandis*	*Pestalotia* sp.	-	264	264	14	3,704
	Shorea robusta	*Lasiodiplodia theobromae*	-	130	130	14	1,720
E. denticollis	*Tectona grandis*	*Pestalotia* sp.	-	46	46	40	1,840
	Areca catechu	*Pestalotia* sp.	-	46	46	40	1,840
Loyolaia indica	dry *Panicum* grass clumps	*Fusarium oxysporum*	-	47600	47600	4	1,90,400

* Mean of ten counts; IM—Immature spores; M—Mature spores.

many thrips, besides Brazil, Guatemala, Costa Rica etc. Studies on litter-infesting thrips at a global level has been documented in a scattered fashion, but without identification of the fungi fed upon. The works of Ananthakrishnan (1983, 1984, 1990), Ananthakrishnan *et al.* (1984), Mound (1974, 1976, 1977), and Mound and Palmer (1982) relating to mycophagous thrips inhabiting litter from the Nearctic, Australia, Japan and Indo-Malayan regions offer useful information. Some of the common thrips from India and associated fungi in litter are:

Mycetophagous species	**Fungus**
Apelaunothrips indicus	*Aspergillus niger, A. flavus and Penicillium*
A. madrasensis	*A. niger, A. flavus and Penicillium*
Ecacanthothrips tibialis	*Aspergillus, Ganoderma*
Stigmothrips limpidus	*A. niger, A. flavus, Penicillium, Pithomyces*
Azaleothrips amabilis	*Aspergillus, Penicillium*
Hoplandrothrips flavipes	*Dreschlera, Pestalotia, Aspergillus*
Sporophagous species	
Bactrothrips idolomorphus	*Pestalotia, Lasiodiplodia,*
Dinothrips sumatrensis	*Lasiodiplodia*
Elaphrothrips denticollis	*Pestalotia, Lasiodiplodia, Phomopsis tectonae*
E. procer	*Pestalotia, Lasiodiplodia*
Neurothrips indicus	*Aspergillus, Penicillium*
Priesneriana kabandha	*Cytospora*

Association of thrips with plant pathogenic fungi is an important aspect demanding increased attention both in view of the number of species associated as well as the number of individuals. In view of the abundance of many sporophagous species in more humid situations in the tropical rain forests, their tendency to aggregate in the larval and adult conditions or both, comprising larger number of individuals in a given unit area, and their feeding pattern variability in terms of host preferences, it may be presumed that their potential to effectively consume large quantities of spores of plant pathogenic fungi is considerable. An equally important aspect is the dual role played by many of these thrips species, namely as mechanical vectors as well as efficient spore feeders of plant pathogenic fungi (Table 8).

5

Diversity of Litter Microarthropods: A Bioecological Assessment

The natural forest litter ecosystem with its floral diversity and specialized microniches supports a variety of macro and microarthropods involving different trophic levels. Diversity of species is not only determined by the type of litter, but also as a function of the complex biotic and abiotic components coupled with the heterogeneity of the litter. Added to this, the chemical nature of litter involving the inorganic components as well as the phenolics, volatiles and flavonoids also contribute considerably to the degree of complexity of the litter insect community. While both species richness and abundance are important aspects of biodiversity, the degree of intraspecific diversity notably of the fungal feeders such as thrips are aspects unique to litter. This chapter is essentially concerned with bioecological aspects of some of the major components of litter.

Species richness is high in natural litter of tropical communities, the microarthropods population varying in direct relation to the amount of litter even in monoculture forests such as those of *Shorea robusta, Tectona grandis*, and *Eucalyptus*. As the litter layer becomes deeper, the successional seres are very distinct and as the litter undergoes an orderly and sequential breakdown, the microfloral and faunal assemblages tend to succeed one another. In this process, such factors as depth and age of litter appear to influence litter communities. Habitat heterogeneity is greater in tropics as also the stability of the niche, the diversity within a community being equal to the total number of species divided by the average number per sample (Allen 1977; Anderson 1978).

Biodiversity of arthropods can be related to the morphology of litter breakdown which is also a useful indicator of the simultaneous chemical change in the faecal pellets contributing to more water retention. Mechanical fragmentation leads to a 15-fold increase in surface area due to microarthropod feeding effects and subsequent chemical breakdown both in the gut and microfaunal assemblages (Gill 1969). Soil arthropods being non-randomly distributed in the soil, the factors causing this non-randomness, presumably relates to the patchy distribution of food or soil water which is the cause of aggregation (Usher 1976; Usher *et al.* 1982).

The Protura, Diplura and Collembola are the major soil dwellers. Japygids like *Anajapyx, Japyx* and *Campodea* are often found in small numbers in moist soils under stones and in humus and litter layers. Proturans are more abundant than Diplura and very common in moist forest and grassland soils which abound in species of *Eosentomon, Acerentomon* and *Acerentulus.* Of all the soil microarthropods, Collembola are the most predominant and stand comparison only with mites. The Onychuridae, Isotomidae, Poduridae and Entomobrydae are richly represented in the soil both in number and species composition. The type of soil, the composition of microflora, soil moisture. and the type of vegetation influence the abundance of Collembola, although variation in pore size and organic content equally influence their variation in density in different soil types (Hale 1965, 1967) Large forms like *Tomoceros, Entomobrya* and *Orchesella* abound in surface litter, while the smaller *Onychiurus, Tullbergia,* etc., occur in deeper layers. In the top 10 to 15 cm Collembola abound in grassland and forest soils, as this is the region where the active breakdown of organic material takes place (Figs. 10, 11). The diameter of the pore cavities in this region is smaller and the moisture content of organic layer is also less. Life forms have been characterised on the basis of distribution of Collembola involving progressive reduction in pigmentation, development of eyes and body size with increasing depth. Surface dwellers which are more vulnerable to predators are *r*-selected animals, whereas soil dwellers tend to be *k*-selected (Slansky 1987). Separation of niches has often been discussed on the basis of the ecological requirements of several species in a community. As a typical example, species of *Tullbergia* have been described along moisture gradients. Similar stratification of species in relation to litter layers L, F and H has been observed in some Collembola in Yercaud (4,500 ft) and has clearly indicated the role of temperature gradients and moisture (Annadurai *et al.* 1988) so that the organization and dynamics of litter soil community form an important area of study.

Soil microarthropods are found distributed in specific patterns, closely related to the status of organic matter and water content (Rusek 1975; Verhoef 1978). Food quality plays a significant role in determining the niche-width of the collembolans (Parkinson *et al.* 1979). As fungal feeders, they contribute to cropping of microbial populations and hence influence metabolic turnover rates (Muraleedharan and Prabhoo 1978). Collembola exhibit phases of active feeding when suitable energy sources are available, followed by non-feeding periods normally

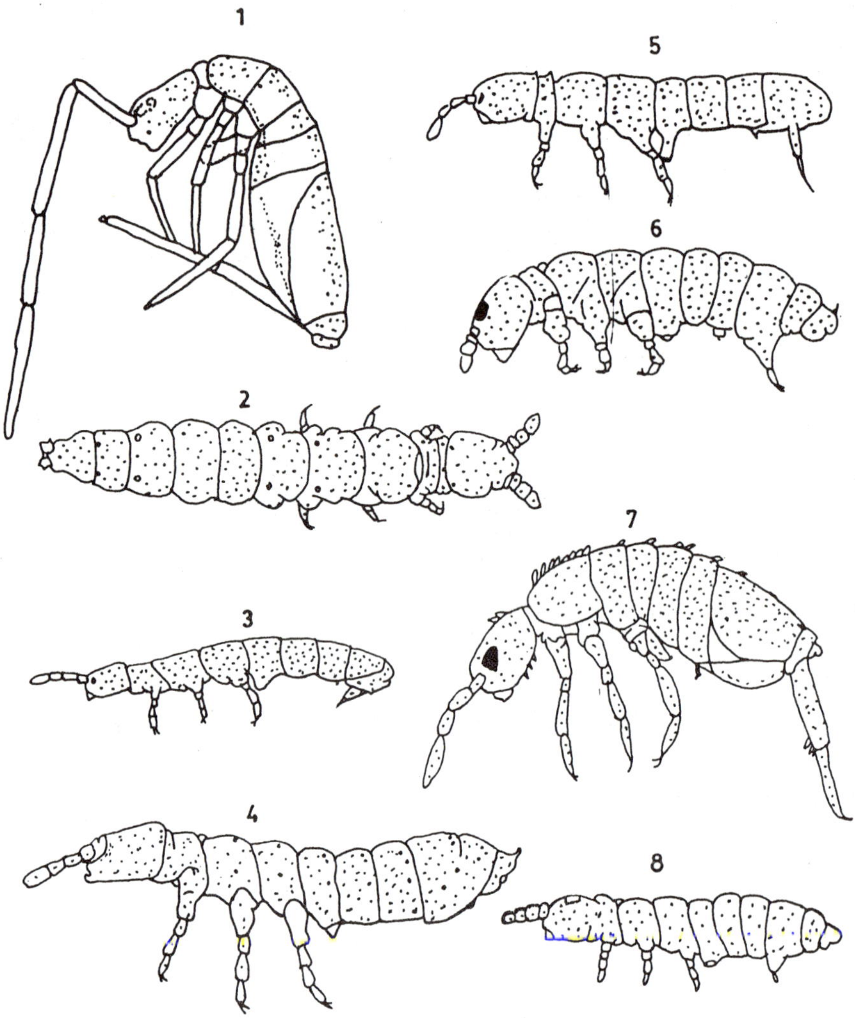

Fig. 10 Some common representative litter Collembola-I.
1. Salina; 2. Tullbergia; 3. Isotomodes; 4. Onychiurus; 5. Folsomia; 6. Hypogastrusa; 7. Lepidocyrtus; 8. Xenylla.

synchronising with moulting and reproduction. Successional patterns are also evident in collembolan assemblages with '*r*' and '*K*' selection strategies (Parr 1978). An example of the '*r*' selection is *Isotoma viridis* and of the '*K*' selection is *Onychiurus ambolans*, wherein the ability of the microarthropod through persistent colonisation and feeding to alter the milieu for its own suitability is considered a major factor (Christiansen 1964; Joose 1981).

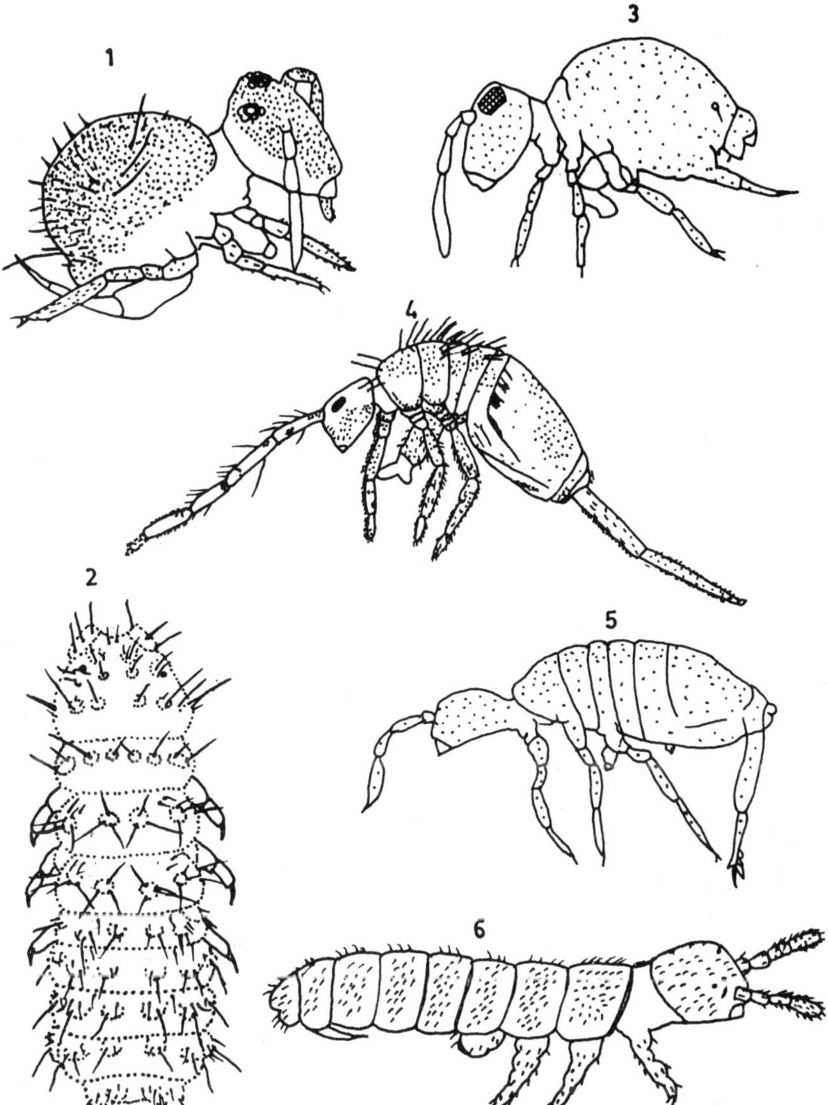

Fig. 11 Some common representative litter Collembola-II.
1. Sphaeridiapumites; 2. Lobella; 3. Sminthurinus; 4. Aphysa; 5. Cyphoderus; 6. Folsomides.

Seasonal Abundance

Seasonal abundance of soil-surface arthropods in *Eucalyptus*-planted area in Warangal (Andhra Pradesh) at MSL (mean sea level) showed that densities of most groups were lowest during summer and highest during rainy season. Formicids like *Monomorium indicum, Crematogaster* sp., *Camponotus* sp., *Pachycondyla* sp. reached maximum densities during rainy season and minimum in winter and summer (Reddy *et al.* 1990). Soil moisture and rainfall were the strongest correlates with densities besides organic matter, soil and air temperature, soil pH, RH and K and P on soil surface. Regular patterns of seasonal abundance were influenced by abiotic factors (Leavings and Windsor 1985). Populations of the total Collembola and constituent species *Lepidocyrtus* sp., *Isotoma trispinata* and *Sphaericia* sp. appeared in large numbers in the early rainy season. *Hypogastruridae* occurred only in the rainy season and decreased subsequently. All the collembolans except *Isotoma* sp. showed a single peak of abundance, i.e. unimodality in the seasonal abundance pattern (Choudhuri and Banerjee 1975). *Isotoma* showed bimodality, with two peaks of abundance, suggesting two generations may occur per year. *Tomoceros minor* was totally restricted to water saturated habitats, inhabiting the humus layer and feeding mostly on fungi and partially decomposed plant materials (Fig. 12).

Collembola attain peak population (131×10^3 m^2) in periods of high moisture in the pine litter during rainfall and species such as *Isotoma trispinata* and *Dicranocentroides salmoni* are the dominant species associated with pine litter (Reddy 1989). Pine plantations in the north-eastern region of India occur in zones of intense rainfall and also where the forest-floor temperatures reach 6°C at the lowest with frost. *Tomocerus flavescens* increases assimilation efficiency when food is in short supply by retaining food in the gut too long (Slansky 1987). *Orchella sineta* is more mesic and can spend portion of its life in unsaturated habitats, allowing it to feed on more undecomposed materials in the upper litter layer (Werner and Dindal 1987). *Folsomia candida* prefers feeding on immature fungal material, though many others feed on more mature fungi. Species with strong ability to decompose chitin exist, because chitin content of fungi increases with age (Slansky 1987). *Cyphoderus javanus* feeds extensively on decaying fungi preferentially often feeding on yeast (Kurup and Prabhoo 1978).

Working on the ecology of Collembola in a deciduous forest floor comprising *Shorea robusta* litter in West Bengal, Hazra (1978b) showed that the patterns of seasonal variation appeared to be different in the different species, with a single population peak. Study of vertical

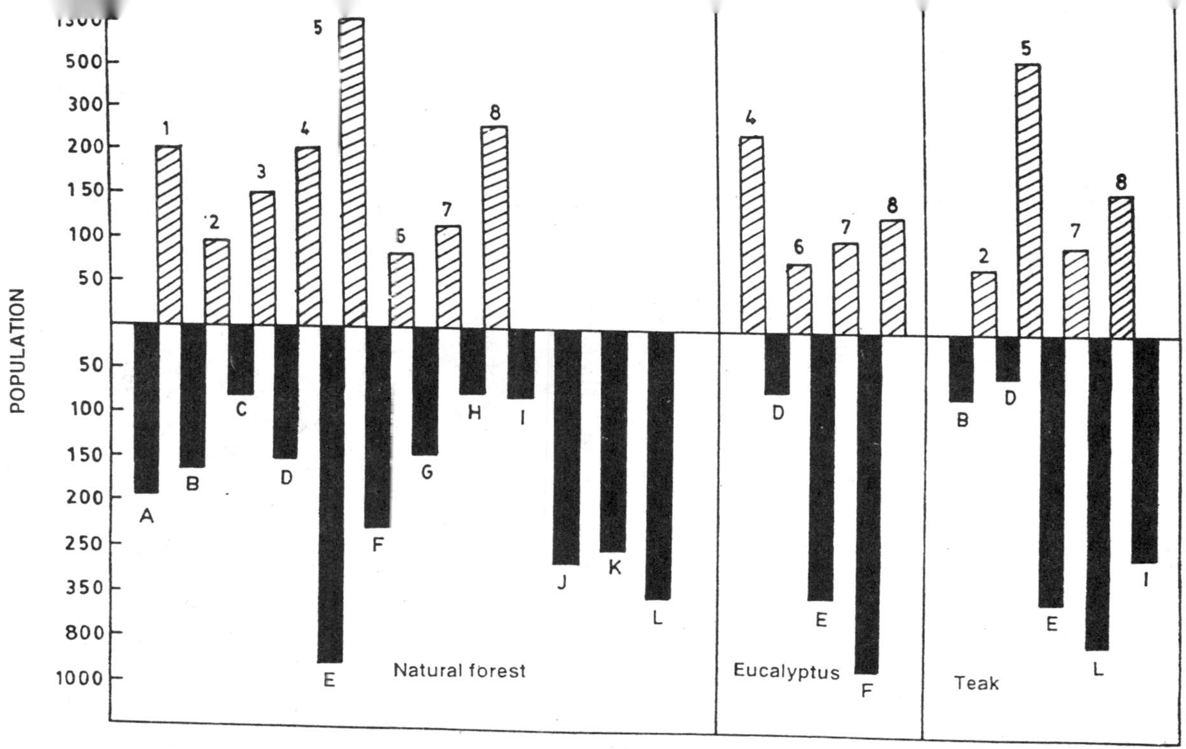

Fig. 12 Mites and collembolans in natural forest and monoculture litter habitats.
(A) *Balaustium* sp. (a), (B) *Balaustium* sp. (b), (C) *E. abinashi*, (D) *Fuscuropoda* sp., (E) *Acarus* sp., (F) *Rhizoglyphus* sp., (G) *Plateremaeus* sp., (H) *Paraxylobates* sp., (I) *Z. saltador*, (J) *Trachygalumna* sp., (K) *A. remota*, (L) *Eupelops* sp.
Collembolans
(1) *A. travancorica*, (2) *S. celebensis* (3) *S. tricolor*, (4) *S. quantturofasciata*, (5) *Xenylla* sp., (6) *Lepidocyrtoides* sp., (7) *Tullbergia* sp., (8) *C. thermophylus*.

distribution revealed that the collembolan population was significantly higher at the surface than lower depth and moisture content of soil had a wide range of variation from 7.1 to 27.5% and this was positively correlated with Collembolan population which was maximum in monsoon months and minimal in summer. The seasonal fluctuation of the dominant species *Cyphoderes, Brachystomella, Lobella, Lepidocyrtus* and *Folsomoides* is very typical. As for the vertical distribution and mean population of Collembola obtained from 0–3 m, 3–6 m and 6–9 m it was evident that Collembola inhabiting the uppermost layers contributed 59.3% with the minimal population occurring in the lowest layer. What is particularly striking is that the species component of Collembola differs with different sites, whether it be tropical deciduous forest or tropical rain forest and whether these are lowland or montane. In this connection a comparison is provided in Table 9, regarding the species composition in a tropical rain forest, and tropical deciduous forest both at altitudes and lowlands (Figs. 13, 14).

Several species of mycophagous thrips inhabit fungus-infested surface litter in natural forests wherein there is a rich accumulation of litter. Both mycelial-feeding phlaeothripine species as well as spore-feeding idolothripids are common. However, their distribution is very restricted in monoculture species such as litter of *Tectona grandis* and *Eucalyptus*. By distribution they are known to occur in tropical rain forests and have been collected from leaf litter in the Australian rain forests, though they are more common in the wet tropics, becoming very abundant in the Western Ghats and forests of north-eastern India, not to mention the other tropical situations met with in S. America and Africa. *Merothrips* and *Erotidothrips* associated with litter and dead twigs in the Tenmalai regions of the Western Ghats are fast disappearing. Species of such genera as *Hoplandrothrips, Hoplothrips, Ecacanthothrips, Adraneothrips, Apelaunothrips, Elaphrothrips, Bactrothrips, Nesothrips* and *Azaleothrips* are very typical of habitats such as drying dead, fungus-infested leaves and twigs. Diverse species of urothripids which are among the smallest of thrips are typical inhabitants of dead and dry leaves and twigs. Many mycophagous species are associated with particular types of fungus infestation in leaf litter and some of the community associated fungi are species of *Aspergillus, Pestalozia, Lasiodiplodia, Cytospora, Phomopsis tectonae and Fusarium oxysporum*. Interestingly enough, instances are known where one species may feed on specific spores of a fungus, or may be polyphagous feeding on a wide variety of spores. Several species are known which feed on spores only of one fungus like *Lasiodiplodia.*

Most of the populations of these litter-inhabiting mycophagous species show phenotypic plasticity resulting in a variety of phenotypes

Table 9. Entomofauna of natural and monoculture forest litter ecosystem

Natural forest-floor litter species	Monoculture forest litter
ACARINA	
Acarus sp.	*Acarus* sp.
Apoplophora	*Balaustium* sp.
Arcoppia sp.	*Dolichedemaeus* sp.
Balaustium sp.	*Eupelops* sp.
Ceratoppia sp.	*Eurotocepheus* sp.
Epicrosejus abinashi	*Fuscuropoda* sp.
Eupelops acromis sp.	*Galumna* sp.
Eurotocepheus sp.	*Lamellobates* sp.
Fuscuropoda sp.	*Rhizoglyphus* sp.
Galumna sp.	*Uropoda* sp.
Hypoaspis sp.	*Zetorchestes saltator*
Lamellobates sp.	
Macrocheles sp.	
Magalotocepheus sp.	
Paraxylobates imitans	
Phyllhermaria sp.	
Plateremaeus sp.	
Rhynchoribates sp.	
Rhyzoglyphus sp.	
Scheloribates sp.	
Setoxylobates sp.	
Trachygalumna sp.	
Uropoda sp.	
Zetorchestes saltator	
COLLEMBOLA	
Brachystomella terrefolia	*Cyphoderus* sp.
Callyntura ceylonica	*Folsomia* sp.
Callyntra sp.	*Hypogastrura* sp.
Cryptophygus sp.	*Isotomodes* sp.
C. thermophylus	*Lobella* sp.
Cyphoderus sp.	*Onychiurus* sp.
Entomobrya sp.	*Onychiurus* sp.
Folsomia sp.	*Salina* sp.
Folsomides sp.	*Sminthurinus* sp.
Hypogastrura communis	*Sphaeridia* sp.
Hypogastrura sp.	*Tullbergia*
Isotomodes sp.	*Xenylla* sp.
Lepidocyrtus sp.	
Lobella sp.	
Onychiurus sp.	

Natural forest-floor litter species	Monoculture forest litter
Paratullbergia indica	
Salina indica	
S. quatturofasciata	
S. tricolor	
Salmon	
Sinella sp.	
Sminthurinus sp.	
Sphaeridia sp.	
Tullbergia sp.	
Xenylla sp.	
DERMAPTERA	
Dicrana kallipyga	*Euborellia* sp.
Eparchus insignis	*Earchius* sp.
Euborellia annulipes	*Labidura* sp.
Forcipula lurida	
F. quadrispinosa	
Forminalabis sisera	
Gonolabidura nathani	
Labia minor	
Labidura riparia pallas	
Metisolabis bifoveolate	
Psalis castetsi	
Pygidicrana eximia	
P. picta	
Pygidicrana valida	
THRIPS	
Apelaunothrips madrasensis	*Azaleothrips amabilis*
Azaleothrips amabilis	*Ecacanthothrips tibialis*
Bournieria indica	*Hoplandrothrips flavipes*
Dinothrips sumatrensis	
Ecacanthothrips tibialis	
Elaphrothrips denticollis	
Gastrothrips falcatus	
Hoplandrothrips flavipes	
Hoplothrips fungosus	
Nesothrips acuticornis	
Neurothrips indicus	
Priesneriana kabandha	
Stictothrips fimbriata	
Stigmothrips limpidus	

Natural forest-floor litter species	Monoculture forest litter
REDUVIIDS	
A. coprolagus	*Pirates affinis*
A. quinque spinosa	*Rhinocoris* sp.
Acanthaspis pedestris	
Catamiarus brevipennis	
E. ochropterus	
Echtrechotes pilicornis	
Ectomocoris tibialis	
Haematorhophus nigroviolaceous	
Lisarda annulosa	
Lophocephala guerinii	
P. mundulus	
Pirates affinis	
Rhaphidosoma atkinsoni	
Rhinocoris marginatus	
ANTHOCORIDS	
Amphiareus constrictus	*Anthocoris* sp.
Anthocoris sp.	*Orius maxidentex*
Buchananiella carayoni	*Xylocoris* sp.
Cardtiastethus nagadiraja	
Cardiastethus sp.	
Cardiostethus pygmaeus pauliani	
Orius maxidentax	
Physopleurella ananthakrishnani	
Scoloposcelis asiaticus	
S. parallelus	
Scoloposcelis sp.	
X. clarus carayon	
Xylocoris (*Proxylocoris*) *distanti*	
LYGAEIDS	
Geocoris ochropterus	Nil
Geocoris sp.	
Graptostethus servus	
Lygaeus sp.	
Metochus uniguttatus	
Naphilus dilutus	
BLATTIDS	
Blatella germanica	
Blattella sp.	
Theria sp.	

Natural forest-floor litter species	Monoculture forest litter
GRYLLIDS	
Gryllus sp.	
COLEOPTERA	
Anthicidae	Carabidae
Bostrichidae	Scarabaeidae
Carabidae	Staphylinidae
Coccinellidae	
Cucujidae	
Hydrophilidae	
Hydroscaphidae	
Nitidulidae	
Scarabaeidae	
Staphylinidae	
Tenebrionidae	
CHELONETHI	
Amblyolpium sp.	*Lamprochernes* sp.
Calocheiridius sp.	*Tyrrannocthonius* sp.
Geogarypus sp.	
Lamprochernes sp.	
Lehitia sp.	
Olpium sp.	
Stenatemnus sp.	
Tyrrannochthonius sp.	

and such phenotypic variability enabled variable exploitation of food resources resulting in the increase of breeding potential. Behaviourally, some species tend to aggregate as in *Ecacanthothrips tibialis* which has a fecundity range of 25–40 eggs/female, whereas in non-aggregating species like *Azaleothrips amabilis* it is 10–16 eggs. There is considerable variation in the egg laying behaviour (Ananthakrishnan *et al.* 1985) from laying eggs in a scattered fashion, to aggregations. *Elaphrothrips denticollis*, common in surface litter of *Tectona grandis*, lays eggs in a horizontal cluster (Ananthakrishnan 1990) (Table 10). *Priesneriana kabandha* in *Eucalyptus* occurs as a dense aggregation laying 10–110 eggs or can be non-aggregated laying isolated eggs. Equally interesting is the fact that *Elaphrothrips denticollis* lays eggs in loose clusters when on *Areca* litter, while it is in dense compact clusters when occurring on *Tectona grandis* (Ananthakrishnan *et al.* 1984; Ananthakrishnan 1990). So environment control, fungal food availability and aggregation behaviour either independently of in combination appear

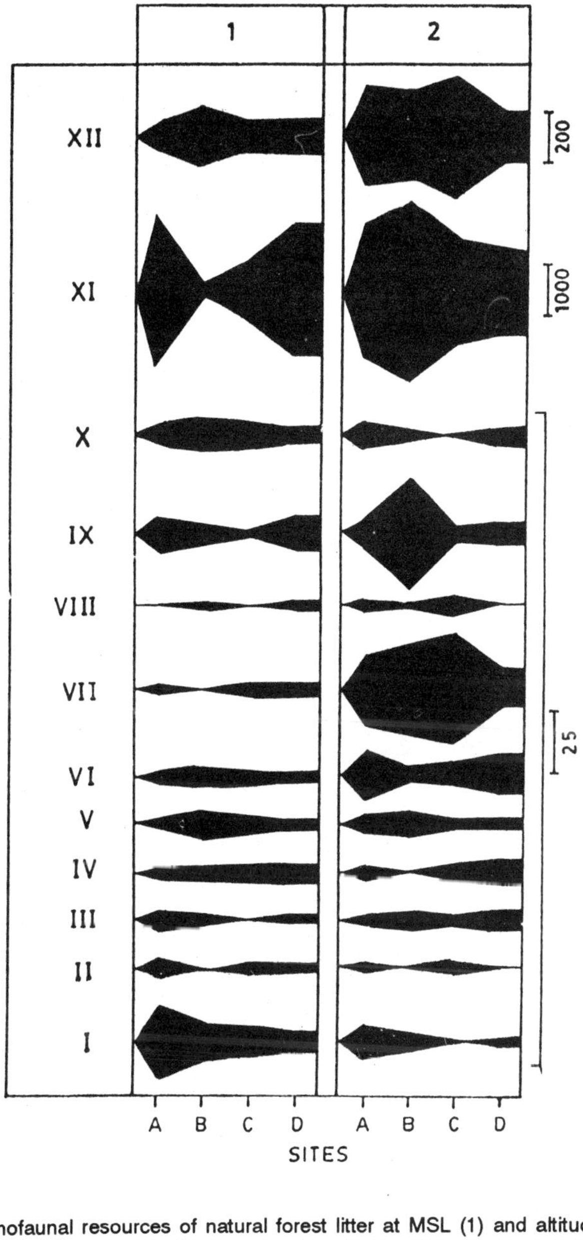

Fig. 13 Entomofaunal resources of natural forest litter at MSL (1) and altitude (2).
I. Blattids, II. Reduviids, III. Scarabaeids, IV. Carabids, V. Staphylinids, VI. Dermapterans, VII. Psocids, VIII. Anthocorids, IX. Lygaeids, X. Thrips, XI. Collembolans, XII. Mites.
The scale represents number of insects.

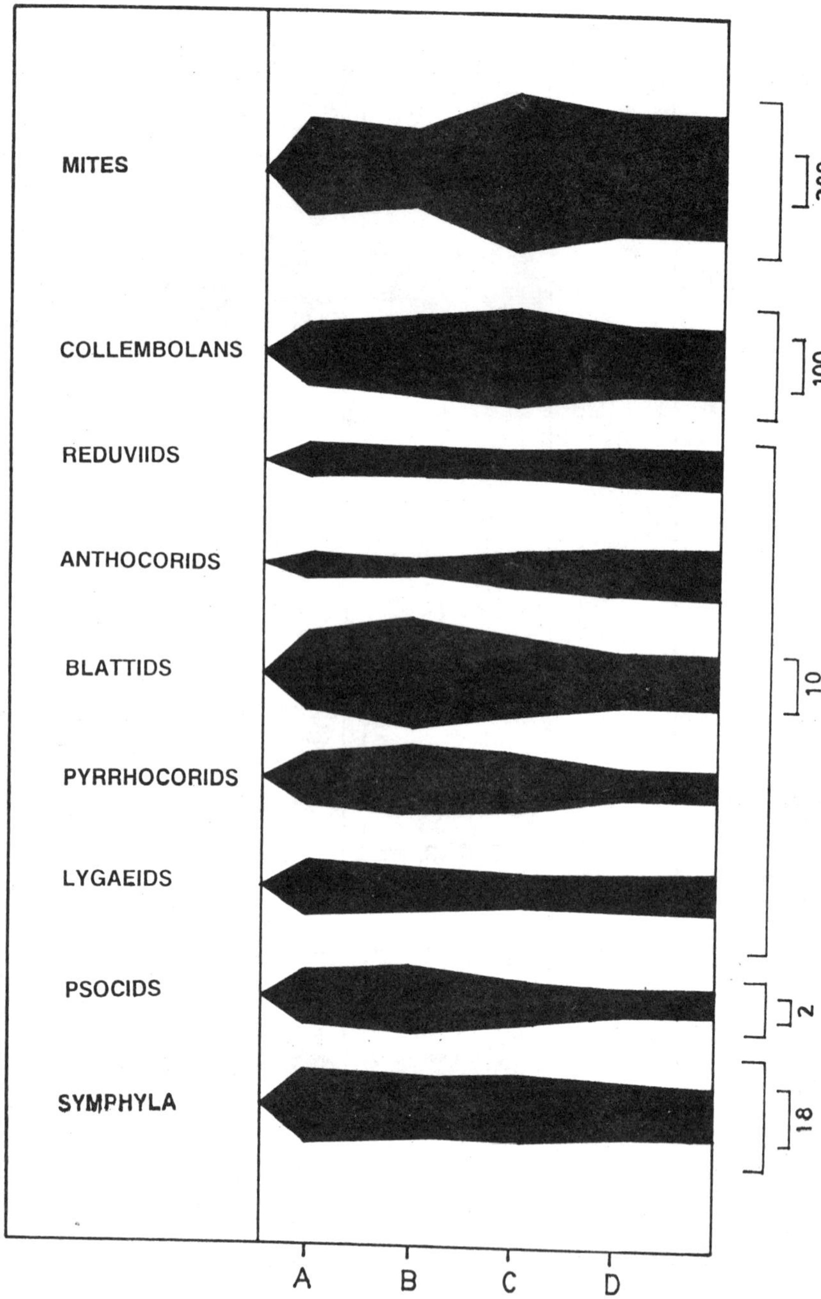

Fig. 14 Entomofaunal resources of natural forest litter of four different sites (A), (B), (C), (D) of Western Ghats (Thekkadi).

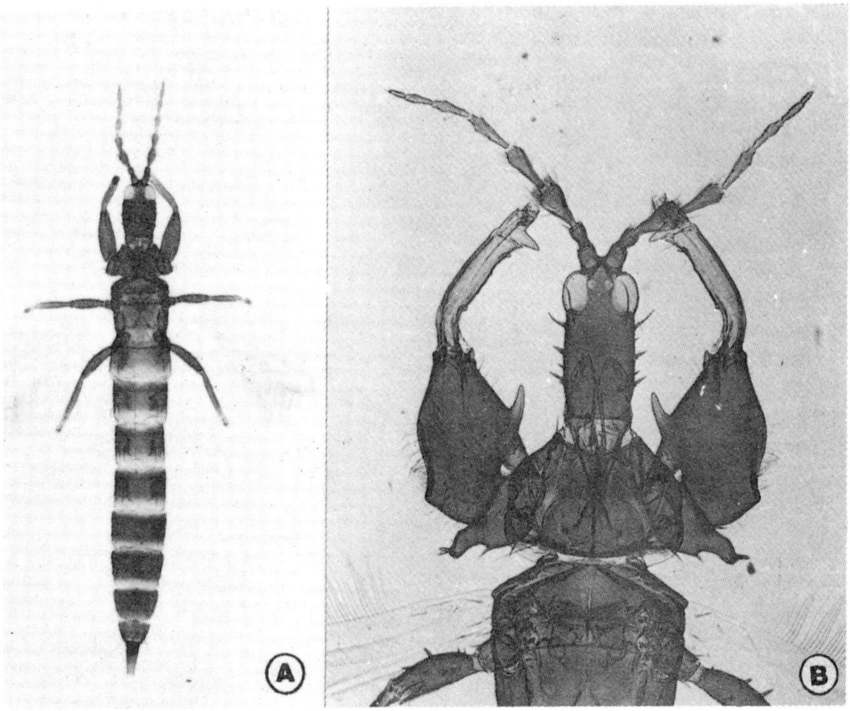

Fig. 15 Phenotypic plasticity in *Ecacanthothrips tibialis*.
(A) Gynaecoid male, (B) Oedymerous male

to trigger reproductive polymorphism, involving oviparity, ovoviviparity and viviparity (Ananthakrishnan 1990). Litter-infesting species like *Bactrothrips idolomorphus*, under field temperatures of 28°C and 92% RH, the oviparous forms laid 27 eggs; at 25°C and 90% RH, 36 eggs; but at 30°C and 80% RH they reproduced both by oviparity and viviparity. The incidence of ovoviviparity as well as viviparity at higher temperatures and RH seems to be adaptation for survival (Ananthakrishnan *et al.* 1984). It is clearly evident that the varying grades of sex-limited polymorphism greatly influence mating behaviour and sexual selection. All these polymorphs are reproductivity compatible and can successfully propagate the progeny, but with varying reproductive rates (Ananthakrishnan 1990).

Others like Dermaptera, Psocids, Dictyoptera, Isoptera, Coleoptera, a few Hymenoptera and some Diptera also occur in the soil, sometimes as juveniles. The termites such as *Reticulotermes, Odontotermes* are important soil dwellers of the tropics and play an important role in the breakup of organic materials and their mixing up with the minerals of

Table 10. Comparative biology of some Mycophagous Thysanoptera inhabiting litter

Thrips species	Fecundity rate	Incubation period (days)	Total duration of life-cycle	Type of reproduction
PHLAEOTHRIPINAE				
Azaleothrips amabilis	20–30 (25)	6–8 (7)	14–19 (16.5)	Oviparity
Ecacanthothrips tibialis	16–24 (20)	3–5 (4)	16–23 (19.5)	Oviparity
Hoplandrothrips flavipes	–	5–7 (6)	22–30 (26)	Oviparity
Stigmothrips limpidus	2–5 (3.5)	2–7 (4.5)	12–24 (18)	Oviparity
IDOLOTHRIPINAE				
Bactrothrips idolomorphus	21–36 (31.5)	7–11 (9)	17–28 (22.5)	Oviparity, Ovoviviparity and Viviparity
Elaphrothrips denticollis	25–28 (26.5)	10–12 (11)	25–29 (27)	Oviparity, Ovoviviparity and Viviparity
Elaphrothrips procer	3–10 (6.5)	1	15–23 (19)	Oviparity, Ovoviviparity and Viviparity
Priesneriana kabandha	10–115 (62.5)	3–4 (3.5)	14–18 (16)	Oviparity

soils. The ability of termites to digest cellulose and hemicellulosic polysaccharides in ingesting litter, is a significant factor contributing to their importance as constituents of litter and soil fauna, enabling rapid turnover of organic matter in ecosystems (Lee and Wood 1971). While termites grouped under the families Mastotermitidae, Kalotermitidae, Hodotermitidae, Rhinotermitidae and Servitermitidae are considered primitive in terms of social organisation and association with symbiotic gut-inhabiting protozoa, termites of the family Termitidae contain bacteria in their gut aiding in digestion and have elaborate social systems.

Though the diversity of termites is generally richest in the tropical rain forests (Lee and Wood 1968) the level of activity appears to be highest in the deciduous woodlands. Several species of fungi are associated with the mounds promoting the decomposition and release of simplified compounds from the mound matrix contributed by dead and decaying wood and partly comminuted litter and debris of plant origin. While fungi through enzymatic degradation release celluloses from lignin-cellulose complexes, the termites subsequently digest the hemicellulose polysaccharides.

Younger populations of termites coincide with monsoon periods. The fecund females lay hundreds of eggs every day, the maximum potential of which is realised in about four to five years of the active reproductive phase. While the females of the primitive order Kalotermitidae lay about 250 eggs/yr with incubation periods of 50–60 days, records of several thousands of eggs per year are available in the case of Rhinotermitidae and Termitidae. Foraging on leaf litter, bark or dead wood is mainly during nights and the intensity varies with the seasons. *Microcerotermes* and *Nasutitermes* are examples of termites that are strictly associated only with wood and dry vegetation. Six species of *Coptotermes* endemic to Australia and *N. exitiosus* exhibit a very high degree of specificity, in that they are associated only with *Eucalyptus* plantations, occurring in densities of nearly 600 individuals/m^2 (Lee and Wood 1968). Nothing much is known about the bioecological aspects of the scavenging termities belonging to Termitidae that feed extensively on freshly fallen litter, except that vegetation is a vital factor determining their distribution.

The lignin-cellulose ratios in the fungus combs are similar to what is assessed in fresh litters and Maldagive (1964) suggests that nearly 50% of litter in forests is consumed by termites, supported by Abe's (1982) study on termites associated with litter in tropical rain forests. Like earthworms, termites literally eat their way into the soil. They are not able to live on fresh litter, but depend entirely on partially decomposed plant material and make a network of underground passages and are humus feeders, transferring the organic contents to soil (Rajagopal and Veeresh 1981). A thorough mixture of organic and inorganic compounds is achieved to form clay-humus complex. The humus component or droppings of termites is more stable than that of the organic horizon of the uppermost soil layer. Intense activity of myriads of termites and ants which rapidly comminute in the litter enabling ready attack by microarthropods.

Among the Coleoptera, the carabids and staphylinids and to some extent coccinellids and tiger beetles are associated with soils principally along with a few others, like the larvae of click beetles or elaterids. Lavae of many of these beetles are subterranean, while the adults contribute significantly to the decomposition of organic matter by their litter feeding or xylophagous habits. The cockchafer beetles or scarabaeids have so wide a range of feeding habits that some of them are important members of soil community along with Elateridae, Carabidae and Staphylinidae. Larvae of nematoceran Diptera like those of tipulids, mycetophilids, the brachyceran tabanids, asilids and empidids, and the Cyclomorpha represented by the muscids, phorids often abound in the soil. Among the Hymenoptera, ants are the most important soil dwelling forms though a few bees and mutillid wasps are also known to inhabit soils. Representatives of *Camponotus, Crematogaster, Formica* are wood dwelling and many members of Myrmicinae build mounds. The availability of food, the physical nature of the substratum as well as moisture content of the soil are important factors limiting the distribution of ants.

Among the soil arachnids, mites are the most predominant. Saprophagous, predatory and phytophagous mites are common and very much involved in the process of organic decomposition and its resultant processes. The mites are the most abundant in organic soil and are represented by the cryptostigmatid detritus feeders; the mesostigmatid and prostigmatid mites include predators and detritus feeders. The Astigmata generally occur in pasture soils. Mites flourish in moist organic soils and forms like *Galumna, Cepheus, Homorobates* have been recorded as inhabiting exposed situations such as in lichens, mosses, etc. The detritus-feeding mites have been known to withstand varying conditions of environment, such as moisture content and pH. They are often rare in cultivated soils and generally prefer undisturbed surface organic layers. The larger forms like *Macrochaetes, Pergemasus* and *Steganocarus* are generally common in surface litter, whereas the medium-sized and smaller forms like *Scheloribates* and *Brachychthomus* penetrate into humus and even deeper down. Immature mites are known to frequent the rhizosphere around plant roots. Vertical migrations are also common in response to changing complexion of the moisture of the litter. Some of the dominant microarthropods in forest and cultivated soils in India are also indicated in Ananthakrishnan (1979) and Prabhoo (1967) wherein a picture of the population size as well as the faunal composition of selected groups of mesofauna is presented (Fig. 16).

The abundance of total Acarina and constituent species such as *Cosmolaelaps* sp., *Stigmaeus* sp., *Lancetoppa willmanni, Eremulus avenifer, Scheloribates praeincisus interuptus, Annectacarus longisetosus, Lamellobates palustris* and *S. fimbriatoides* were seen to be in maximum number during the rainy season. Peaks varied with sites and places (Choudhuri and Banerjee 1975), a single peak of abundance of acarines being in July–August, and least during May.

Within a forest complex gradients exists so that the soil is usually pictured as a 'system of habitat mosaics' of various sizes and shapes superimposed both within and upon one another (Mitchell 1978). Among these mosaics various gradients such as temperature and moisture are evident along which species orient themselves. Various species of *Scheloribates* and *Oppia* orient themselves differently in response to such gradients. These oribatids exhibit wide morphological and biological spectrum enabling them to occupy a variety of such ecotopes. Each species has its own distinct spatial distribution, often complicated by several fluctuations and differences between development

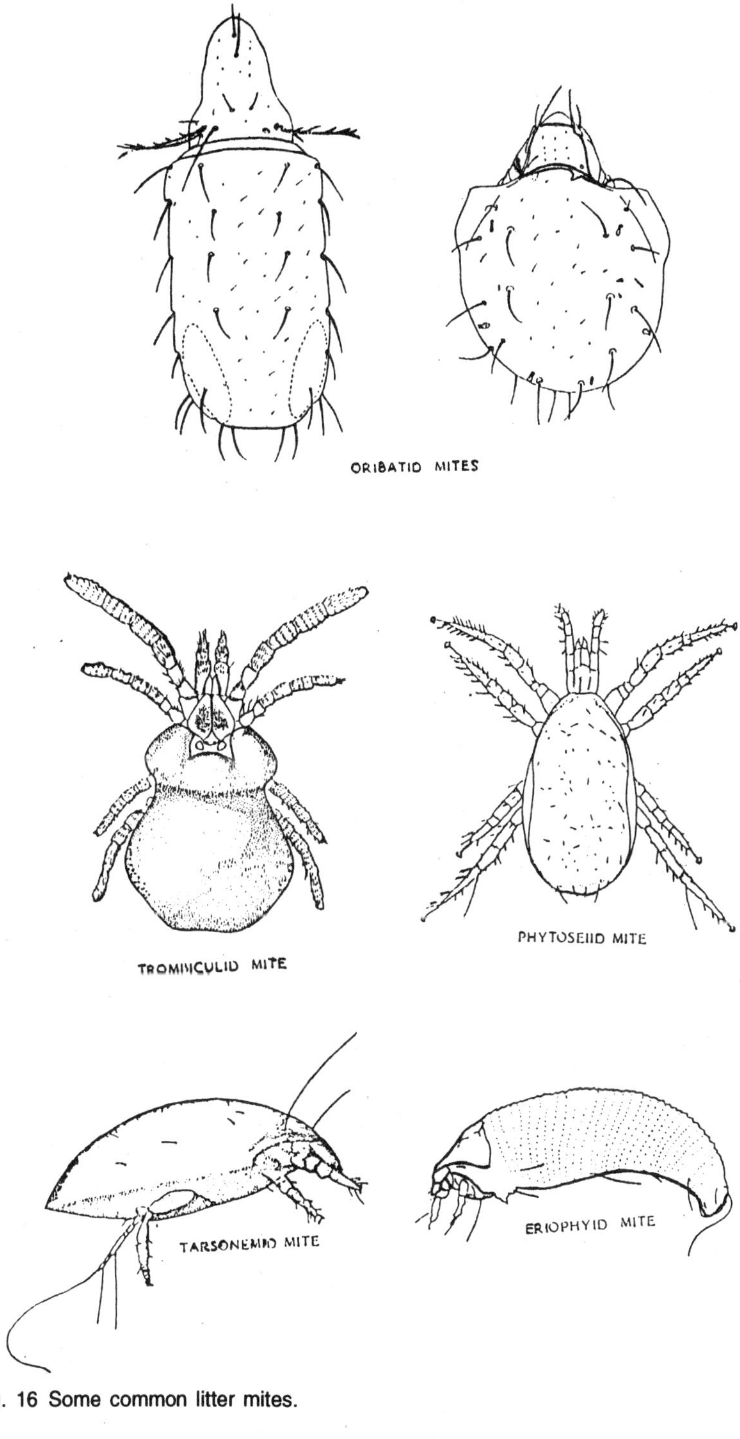

Fig. 16 Some common litter mites.

stages (Aoki 1967). Distinct horizontal and vertical distribution patterns are evident in oribatids and other mites with striking differences in the L layer inhabitants and F-H inhabitants. Depth of organic matter is positively correlated with only the F-H inhabitants. Mitchell (1978) has shown a contrast between L, F and H inhabitants and only the F-H ones tend to show a significant positive correlation with organic matter depth. L-inhabitants are not sensitive to moisture as the H-F ones which shows seasonal variations in distribution with a downward vertical movement varying with species.

Oribatids are the most abundant soil mites having three feeding types—macrophytophagous, microphytophagous and non-specialists or panphytophagous (Luxton 1972). Among the fungal-feeding microphytophagous species, there are obligatory mycetophiles and mycetobionts. Only the micro and panphytophagous species possess enzymes necessary to digest and assimilate fungal material, other accidentally feed upon fungal tissues and these are dispensed as fungal propagules by way of faecal pellets. Species of fungi such as those of *Phoma, Poecilomyces, Penicillium* and *Trichoderma* are seen in the gut of mites. Inhabitants of L layer consume higher proportion of pigmented hyphae than F-H layers. Various fungal species affect the reproductive success of oribatids indicating the nutritional quality of the fungal material which may affect the overall density and structure of the mite community allowing those species which can utilise the wide range of fungi to be more successful (Mitchell and Parkinson 1976). Some oribatid mites show fungal feeding preferences, selective feeding tending to alter the composition of the microbial community by allowing the less palatable species to colonise more rapidly. Since fungi are responsible for most of the actual metabolic degradation of organic matter in forest soils, any activity which affects the structure of the fungal community either quantitatively or qualitatively may alter the decomposition process. Oribatid mites as predominant inhabitants of forest litter are essentially mycophagous and saprophagous and are intimately associated with decaying vegetation. Oribatid macrophytophagous species have a food assimilation efficiency of 10–15%, as compared with 50–65% for oribatid microphytophagous species (Luxton 1972). The most common oribatid mites are *Galumna* and *Scheloribates* found distributed in the soil and are associated with decaying the decomposing vegetation and 60% the oribatids carry fungal propagules. Patterns of egg laying the frequency vary between the species. For instance, *Rhysotritia duplicata* breeds throughout the year and is not affected by changes in environmental temperature or moisture conditions. On the other hand, *Pergalumna omniphagus* lays eggs only during winter. *Galumna flabellifera* is a frequently encountered species in litter which exhibits a total development duration of 48 to 67 days depending on the nutritional status of the food available (Reddy *et al.* 1978). *Lepidocarus ornatissimus* and *Archegozetes pongisetosus* exhibit contrasting life cycles. While *Lepidocarus ornatissimus* undergoes 5–6 days of incubation in the egg stage, *Archegozetes pongisetosus* exhibits an incubation period of 15–16 days.

Contributions of oribatids to metabolic energy release is small. Macrophagous mites obtain their food mostly by fragmenting dead and higher plant litter thus catalysing primary decomposing activity by increasing the substrate surface area for further microbial growth (Griffith and Bowman 1984). Some possess an array of polysaccharide enzymes. Food bolous studies have shown that in species such as such as *Steganacarus spinosus* most boluses containing microbial material were found in animals extracted from 3–6 cm depth of soil (22.2%), whereas from 0.3 cm depth soil (10.2%) and 8.7% from the litter. As many as 91% of *S. spinosus* in litter contained food, 84% at 0.3 cm, 78% at 3–6 cm, showing that the percentage of oribatids with filled guts varied according to their depth distribution in soil. A closely related species *S. magnus* had values of 66%, 55%, and 20%. It is evident that while this orbitid preferred to inhabit 0.3 cm depth, is appeared to feed mostly in the litter zone.

6

Predator Dynamics in the Litter Environment

The surface layers of litter abound in macropredator species as against the micropredators associated with the F layer. Macropredators include such Hemiptera as reduviids, anthocorids, lygaeids, Coleptera such as carabids, staphylinids, pselaphids, earwigs or dermapterans and the Chelonethi or pseudoscorpions. Predatory Collembola, mites, thrips, comprise the micropredatory elements. Predators encounter several alternate prey in the litter ecosystem, with the advantage that the predators would be less vulnerable to changes in the abundance of any one of them, than a predator with a more restricted diet. The coexistence of a host of collembolans and mite predators is a fine example and from the point of view of larger predators like reduviids, increasing trophic levels sustain their population. The wide choice of food decreases the frequency of explosion and crashes of the predator population. The F layer in litter system being comparatively stable from the viewpoint of the local environment, unlike the surface layer, there is a certain amount of stability of the population of prey species and predators. Species may rise and fall with increasing predation intensity. All the same, the degree of diversity of prey–predator population also depends on the nature of the substrate, with changes on the relative consistency of specific composition of species. Species may experience different spatial heterogeneity, different competition, different predators and different environmental strategies in different seasons. Diverse and complex systems are intrinsically stable and stable communities with three trophic levels can be constituted from predator–prey components (May 1973).

Predation does not act as a population limiter in the same sense as resource shortage. Prey mortality is likely to be lower, since with more prey the number of predators per prey individual will be lower. Thus if a predator population remains constant, mortality would vary inversely with prey density and population instability would be encouraged.

On the other hand, if the predator population rises proportionately more than the prey population, to which it responds, then the predatory element may act as a population limiting factor on prey. Predator intensity as a function of local concentration of prey is more in the litter ecosystem, more because of its heterogeneity. Prey species which increase at the expense of its competitor individuals of the pre-reproductive age, are more prevalent and so young individuals are more prone to predation than mature individuals. As the number of predator species decreases, selection acts strongly in favour of these individuals, providing superior escapability. Predators keep competing populations below their respective carrying capacity, promotes coexistence and predatory biodiversity. Predators do not allow for coexistence, but changes the degree to which competitors depend on each other's fitness. The existence of alternate food will mean that a predator will feed less intensively on the prey. The predator curve is thus less steep. It will vary as the relative abundance of its prey species change. In addition, the predator may shift its preference towards the more common prey species.

Prey–predator system is intensively oscillatory. Such oscillations arise also for time lags and changes in age structure, i.e., increased birth rate of predators responding to high food concentration needs time to be effective on the rate of reproduction. Dispersion pattern of prey is equally important in affecting prey–predator interaction. When prey is scattered with other species, the time and energy expended between food items increases for predation, i.e., time used by predators for search and decision making. Individual predators respond directly to the abundance of their prey through changes in behaviour, but the availability of the prey affects both survival and reproductive success so that predators can also respond indirectly to prey density and increase their numbers through reproduction. *Numerical response* is therefore applied to this change in density of the predators as a result of increase in prey density. Whenever a prey population increases in density, predators also tend to increase in density consuming more prey, resulting in a *functional response.* Generally the density of the prey and the numerical response of the predator bears an intense relationship. Functional responses on the other hand are determined by many factors especially the behaviour of the predators in search of food as well as for detecting the availability of prey. The flow chart provided for the reduviid *Rhinocoris marginatus* explains the different responses.

For continued development, each predator has to seek a particular number of biomass of prey. If the predator feeds on more than one species of prey, changes in the density of one prey species will bring about a shift in the percentage composition in the diet of the predator. Adult fecundity is the main influence on the numerical responses of predator and prey density. True predatory insects mostly ingest a certain biomass of prey above energy maintenance levels to produce eggs and deposit them. This ensures that nymphs and larvae will hatch in the vicinity of the prey. As such, the overall rate of increase under constant environmental conditions are the developmental rate, survival rate of each larval instar, and fecundity of adults. The components change with different predators. Huffaker (1958) highlights four important factors that may constitute to the predator–prey interactions, viz., environmental heterogeneity; spatial relationship between predator and prey; rates of dispersal and influence of change of quality of prey found. Predation plays an important role in energy flow through the litter community. It is a selection factor in nature, particularly in the evolution of their prey and in so doing enables maintenance of a vigorous population of a prey species. Predation pressure is an important evolutionary force, since on it depends the density of insect population they are exposed to and the defensive mechanism that has evolved in the interaction process. Comparative studies of different types of litter ecosystems will throw sufficient light on the influence of predation. Brief accounts of the role of some lygaeid, reduviid, anthocorid, dermapteran, coleopteran, pseudoscorpion and mite predators from litter are provided in this chapter.

Predatory Hemiptera (Fig. 17)

(a) *Geocoris ochropterus* (Lygaeid)

Geocoris ochropterus and *G. bengalensis* are facultative predators, occurring in leaf litter and feeding on a variety of insects like thrips, mites and collembolans. Prey consumption varies in accordance with the size and the nature of the prey. Thrips were found to be the most preferred prey type and the prey consumption under laboratory conditions differed depending on the developmental stage and sex. Species of *Geocoris* complete their life cycle within 25 days and the longevity of the adults varies between 30 and 50 days. The maximum fecundity recorded was 60–66 eggs for *G. ochropterus* when fed on *Ecacanthothrips tibialis*, a common litter thrips. In *G. bengalensis* it is 47–166 eggs (Mukhopadhyay 1988). The pre-oviposition period, longevity and fecundity vary considerably when fed with different thrips species as well as with other groups of insects. The eggs of *G. bengalensis* are cylindrical with considerably tapering ends, more suited for a litter

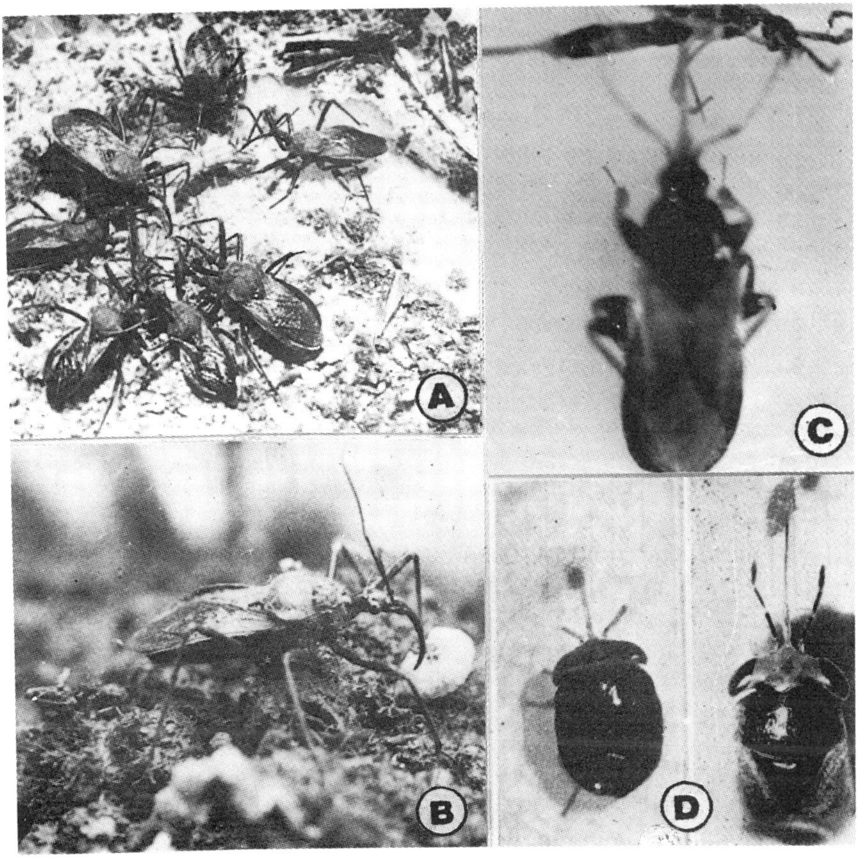

Fig. 17 Some predatory Hemiptera.
(A) Group of reduviids
(B) Reduviid feeding on a beetle larva
(C) Anthocorid feeding on a Collembola
(D) Geocorids

habitat. Eggs are pale, with a tint of pink when fresh, but turning red on maturity. 'Big-eyed' bugs as they are called, they are omnivorous predators playing a useful role in the biological control of insects. The lygaeids *Metochus uniguttanus*, *Graptostethus seruus* and *Naphiellus dilutus* are secondary consumers.

(b) Anthocorids

Anthocorids are efficient predators of thrips inhabiting litter, the well-known species being *Scoloposcelis parallelus* feeding on eggs, larvae and adults of *Ecacanthothrips tibialis*, whereas *Xylocoris clarus* feeds *on Apelaunothrips madrasensis.* The first and second instar nymphs of these species prefer thrips larvae, whereas the later stages including adults do not show any stage preference. During post-embryonic development of these anthocorids, around 70–90 thrips are consumed. The duration of the life cycle from eggs to adult ranges from 28 to 32 days for *Scoloposcelis* and from 32 to 42 days for *Xylocoris.* The maximum number of eggs laid by a female ranges from 40 to 86, the longevity of adults by 59–61 days (Ananthakrishnan 1984).

(c) Reduviids

Reduviid bugs are mostly entomophagous, feeding on the insects of various habitats including the leaf litter ecosystem. The reduviids, *Ectomocoris ochropterus, E. tibialis, Lisarda annulosa, Pirates affinis, Rhinocoris marginatus* and *Acanthaspis pedestris* are some of the key species inhabiting leaf litter ecosystems of natural forests. *E. tibialis* and *E. ochropterus* feed on termites beetle grubs and ants. Species like *P. affinis* predate on carabid beetles *Omphora pilosa* and *O. atrata*, whereas *L. annulosa* hunt for Collembola, thrips and psocids.

The moving prey offers an important stimulus in the feeing behaviour of reduviids, arousing the predator from a state of akinesis to a high level of excitation. On perceiving continued prey movement, the predator orients towards the prey by raising the body and by extending the antennae in front, slowly stalks the prey. The rostrum may be frequently fully extended during pursuit. When the prey is within the striking distance, even without the antennal contact, the forelegs are raised and used to pin down the prey body for subsequent stabbing with the stylets and injection of toxic saliva. The sequence of events exhibited by most species in killing soft-bodied prey may be said to be: arousal-orientation-approach-antennal extension and contact-raising of forelegs-extension of rostrum-pinning down prey with forelegs-insertion of stylets-injection of saliva.

The rostrum of Harpactorinae is long and slender and capable of considerable forward extension. This is aptly suited for the pin and jab type of feeding behaviour, where the rostrum and stylets are used to attack the soft-bodied prey types without the necessity to actively chase, to grab, to hold or to pounce on the prey as done by the chase and pounce type of reduviids. The latter reduviids are characterised by short curved rostrum capable of less than 45° forward extension, with which they probe for suitable site only after establishing a firm grip of the heavily sclerotised prey body.

The sequence of events depicting the predatory behaviour of reduviids such as *Rhinocoris* is provided in Fig. 18.

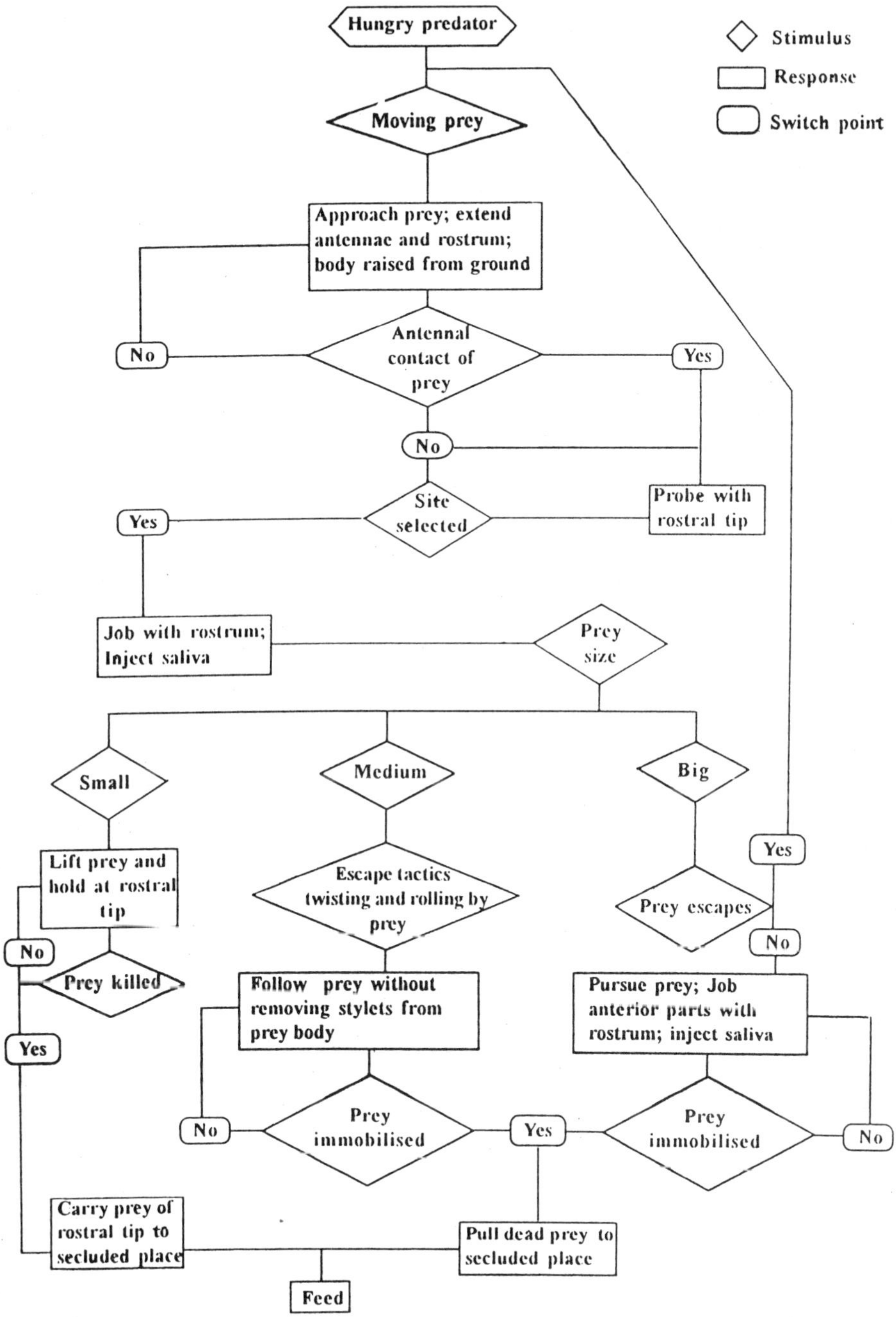

Fig. 18 Flow chart depicting the predatory behaviour of *Rhinocoris marginatus*.

The post-embryonic developmental duration, fecundity, longevity and number of prey consumed during the life time are the important parameters to assess the predatory efficiency of an individual. The predatory efficiency of all the species of reduviids were analysed in the laboratory by rearing them, providing different prey types. The reduviids *E. ochropterus* and *E. tibialis* showed least nymphal duration when fed on gryllids (80.5 and 68.3 days respectively) compared to that of grubs of Coleoptera (103.4 and 85.4 days respectively), (Tables 11, 12). The nymphal duration of *P. affinis* when fed on *O. pilosa* and *O. atrata* was 74.6 and 76.8 days and *A. pedestris* when fed on termites, grubs and ants was 43.8, 67.6 and 66.92 days respectively. The nymphal duration of *L. annulosa* when fed on collembola, psocids and termites was 54.3, 51.8 and 46.6 days respectively. The longevity of *A. pedestris, E. ochropterus, E. tibialis, P. affinis* and *L. annulosa* when reared on different prey types showed very little variation, whereas a distinct variation was noticed in the fecundity of the predators (Fig. 19). *E. ochropterus* and *E. tibialis* when fed on gryllids showed increased fecundity (30 and 326 eggs/female respectively), while *P. affinis* when reared on the carabid beetles *O. pilosa* and *O. atrata* showed a slight variation in fecundity (126 and 114 eggs/female respectively). *A. pedestris* when reared on termites, grubs of Coleoptera and ants laid 237, 158 and 120 eggs/female respectively, whereas *L. annulosa* when fed on collembola, psocids and termites laid 112, 126 and 142 eggs/female respectively.

Table 11. Impact of prey on post-embryonic development longevity and reproduction of *Acanthaspis pedestris*

Post-embryonic development (days)	Prey types		
	Termites	Black ant	Maggots
I Instar duration	7.0 ± 1.0	9.25 ± 1.71	11.0 ± 1.0
II Instar duration	7.5 ± 0.71	12.33 ± 1.53	10.33 ± 1.53
III Instar duration	6.33 ± 1.53	8.50 ± 1.29	6.5 ± 1.29
IV Instar duration	4.50 ± 2.12	18.25 ± 1.71	10.25 ± 1.71
V Instar duration	18.5 ± 2.08	18.6 ± 3.21	29.5 ± 2.65
Total nymphal duration	43.8	66.9	67.6
Adult longevity	121.33 ± 6.51	113.0 ± 19.80	11467 ± 11.15
Preoviposition period	6.67 ± 1.53	13.0 ± 1.21	9.67 ± 1.53
Oviposition period	108.0 ± 3.0	101.0 ± 16.97	105.0 ± 11.79
Post-oviposition period	2.33 ± 0.58	3.67 ± 1.53	3.0 ± 1.41
Total number of eggs	237.33 ± 14.57	120.0 ± 2.83	158.0 ± 29.5
Fecundity index	35.58	9.23	16.34
Incubation period	8.0 ± 1.0	9.66 ± 1.53	10.0 ± 1.0

Table 12. Reproductive biology of *Rhinocoris marginatus*

Developmental stage (days)	Prey types		
	Caterpillar	Grasshopper	Grub
I Instar	6.5 ± 0.5	4.5 ± 0.75	5.5 ± 0.61
II Instar	6.0 ± 0.4	5.0 ± 0.82	7.0 ± 0.42
III Instar	6.0 ± 0.6	6.0 ± 0.64	5.5 ± 0.63
IV Instar	4.5 ± 0.9	7.0 ± 0.43	5.5 ± 0.61
V Instar	10.5 ± 3.45	11.0 ± 2.56	9.5 ± 5.12
Adult female	105.0 ± 26.82	112.0 ± 31.56	85.0 ± 11.7
pre-oviposition period	10.5 ± 2.72	15.0 ± 3.6	21.0 ± 5.27
Oviposition period	89.5 ± 5.30	92.0 ± 6.25	60.0 ± 11.5
No. of egg messes	7	6	4
No. of eggs	225.0 ± 14.85	182.0 ± 15.33	117.0 ± 38.58
Fecundity index. Total no. of eggs laid/pre-oviposition period	21.43	12.13	5.57
Incubation period	8.0 ± 0.8	9.0 ± 0.91	8.5 ± 1.1

The predatory efficiency of *A. pedestris* was high on termites (638/female) as compared to that of grubs (401/female) and ants (109/female). *P. affinis* consumed the carabid beetles *O. pilosa* (32/female) and *O. atrata* (43/female), whereas *E. ochropterus* and *E. tibialis* preyed upon 48 and 46 gryllids/female as well as 95 and 76 grubs/female respectively. *L. annulosa* consumed 330 collembola, 107 psocids and 324 termites/female, whereas the anthocorid *S. parallelus* consumed 216 thrips, 252 mites and 272 Collembola/female respectively during their life time.

(d) Coleoptera

Carabid beetles form one of the most important predator groups on the ground surface exerting a profound influence on the soil invertebrate fauna. Some of the well-known predatory species come under the genera *Carabus, Polydorus, Bembidion, Notophilus* etc. *Polydorus* species common in forest floor, is considered a deep litter form which lives in wet or damp parts of forests, occur in the humus layer at a depth of 3.5–7 m (Adis *et al.* 1948) (Table 13).

Intrinsic stability of the population density of many carabids caused by cannibalism, make them good potential agent to stabilise the density of the population of their prey at low levels. Carabids show a definite numerical response and reaction to the increased density of their prey population. At high density the intensity of their cannibalism among the carabid larvae diminishes and predation becomes more important. Carabids show aggregation when prey species are abundant and this

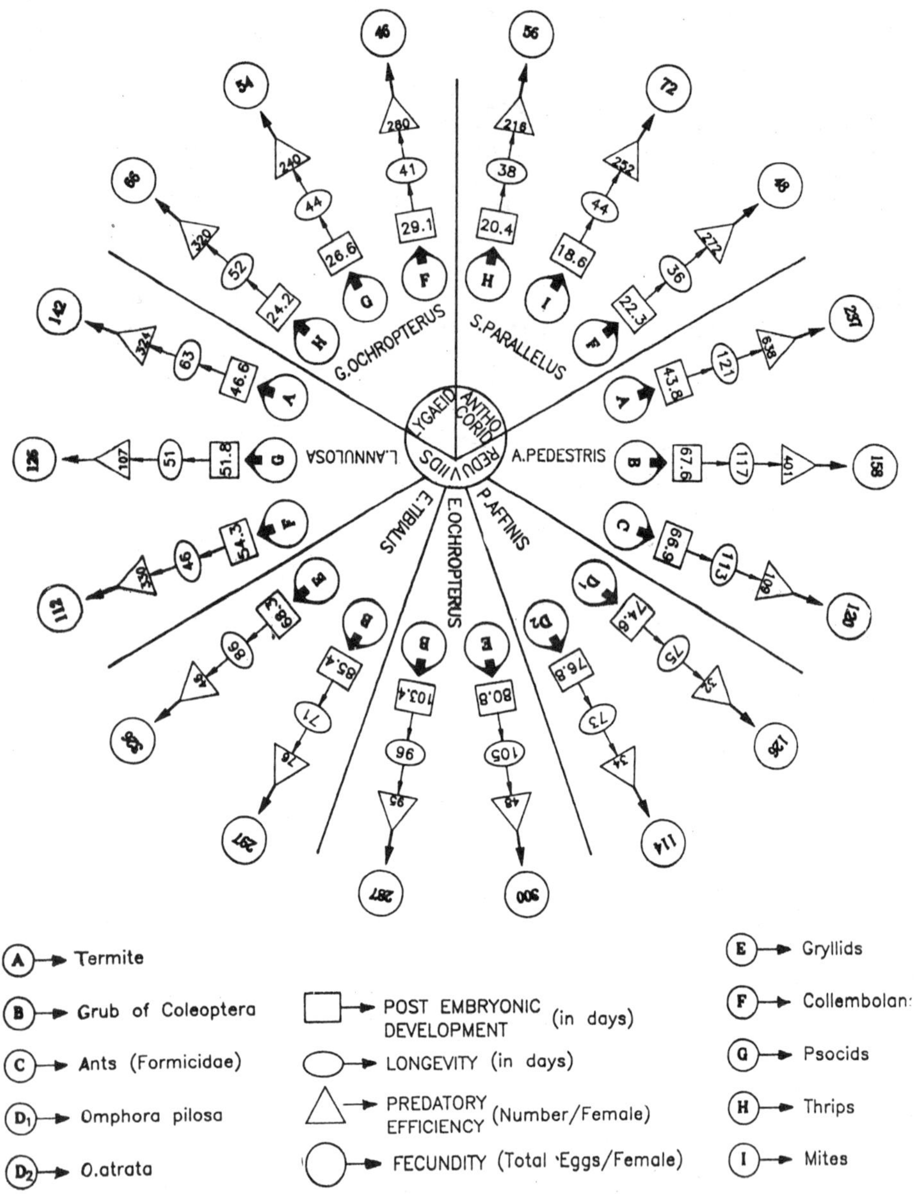

Fig. 19 Comparative biology of some predatory reduviids in the litter ecosystem.

Table 13. Vertical distribution of *Polydorus* in the soil (3.5–14 m depth) (after Adis *et al.*, 1986)

Depth of soil (m)	Soil type	Catch (%)
0–3.5	Upper humus layer marshy litter	27
3.5–7.0	Lower humus layer	55
7.0–10.5	Lower humus layer wet clay sandy soil	14
10.5–14.0	Clay sandy soil	5

provides a potential powerful stabilizing mechanism for predator–prey interactions (Brunsting *et al.* 1986).

Among specialist predators of mites the Scydmaenidae and Pselaphidae are particularly prominent in forest litter, feeding more notably on oribatid mites, which are liable to exert a profound influence over soil invertebrate fauna. Scydmaenids have sharp, sickle-like mandibles to break the hard shell of mites. *Cephennium* feeds exclusively on hard-shelled mites. The number litter of inhabiting staphylinids are many, usually found in the top layer feeding partly as predators and partly as detritivores. Species of *Staphylinus, Stenus, Astillus* predate both in the larval and adult stages. Larvae of Cleridae and Histeridae are also known to be predatory, and they are of comparatively rare occurrence and are known to adopt fairly simple searching strategies. Equally rare are coccinellids whose larvae are often found lying in wait for Collembola.

(e) Dermaptera

Euborellia stahli is a common earwig often found in comparatively high densities of around 45–50/m^2 in *Eucalyptus* litter, being meagre in numbers in teak litter. Eggs are laid in clusters, with an average of 12–8 eggs/female in a cluster with a maximum of 28 and minimum 2. Oviposition occurred almost 15–20 days after mating and the number of eggs laid vary from 21–40. As many as 4 nymphal instars are present. Freshly hatched nymphs are whitish-yellow, turn brown in 4–6 hr, the first 2 instars always stay within the vicinity of the mother. Nymphal instar stage takes 40–50 days, the individual duration of the instars (I–IV) are, 10–25, 6–20, 5–19, 6–15 days respectively. Maternal care is very typical, the female sitting over her eggs guarding them, and if the eggs are separated they do not hatch. If disturbed while brooding, the mother takes the eggs one by one with its mandibles and removes them to a safe corner. If by some reason, the eggs are scattered the mother consumes them.

Species like *Labidura riparia, Euborellia stahli* feed voraciously on various microarthropods present in litter. After the first moult the males and females are cannibalistic and females increase their cannibalistic tendency during their early adult life. *Labidura riparia* has five instars, as against four in *Euborellia.* The nymphal duration in *Labidura* is 25–35 days. The males and females of *L. riparia* have a longevity of 21–152 days in males, 35–176 days in females; and that of *Euborellia* 20–185 days in males and 29–189 days in females (Fig. 20).

(f) Predatory Mites

Among mites, the soil-dwelling Prostigmata and Mesostigmata are reported to be predatory. In prostigmatids the mouth parts are adapted for piercing and sucking and in mesostigmatids for sucking and tearing. The prostigmatid families Bdellidae, Cheyletidae, Rhagiidae and Trombiidae are predators as well as Tydeidae on eggs of small arthropods. The predatory Mesostigmata comprise genera such as *Pergamasus, Macrocheles, Veigaia, Gamasolaelaps* occurring in an important place in the food webs of the litter community. The mite *Pergamasus* is viviparous, being cannibalistic when no other food is available (Wallwork 1970).

Predatory prostigmatids include such species as *Cheyletus eruditus* predatory on other mites, psocids and beetles. *Galumna flatellifera orientalis* is an oribatid with an egg incubation period of four to five days subsequent to which an inactive I instar emerges. With the hardening of cuticle in a couple of hours duration, the young ones commence feeding preferentially on *Alternaria.* The total nymphal duration lasts 15 days. Pergamasids are the largest and fastest moving mesostigmatids in temperate forest leaf litter. *Veigaia* predates on Collembola, and other mesostigmatids. Though occurring only in modest density, they are very heavy predators. The predatory prostigmatids such as bdellid mites attack Collembola, while the erythracids of the genus *Balaustium* feed on scale insects or thrips (Fig. 21).

(g) Chelonethi

Chelonethi or the pseudoscorpions abound in surface litter predating on Collembola, mites and thrips and the more common species include *Tyrranochthonius madrasensis* and *Compsaditha indica.* Mostly, where the females carry egg mass in a thin sac with 4 or 7 eggs, the developmental period from egg extrusion from the ovary to hatching of protonymphs take 9–10 days. In *Lamprochernes indicus* the eggs sac contains 11–13 eggs and the protonymphs hatch takes in 10–12 days, while *Stenatemnus indicus* has 25–27 eggs in the egg sac taking about 75 days for the hatching of protonymphs. *Hygrochelifer indicus* carries 15–17 eggs taking 14–16 days for the protonymphs to hatch. Being predatory in habit, they use their pedipalps for capturing and paralysing

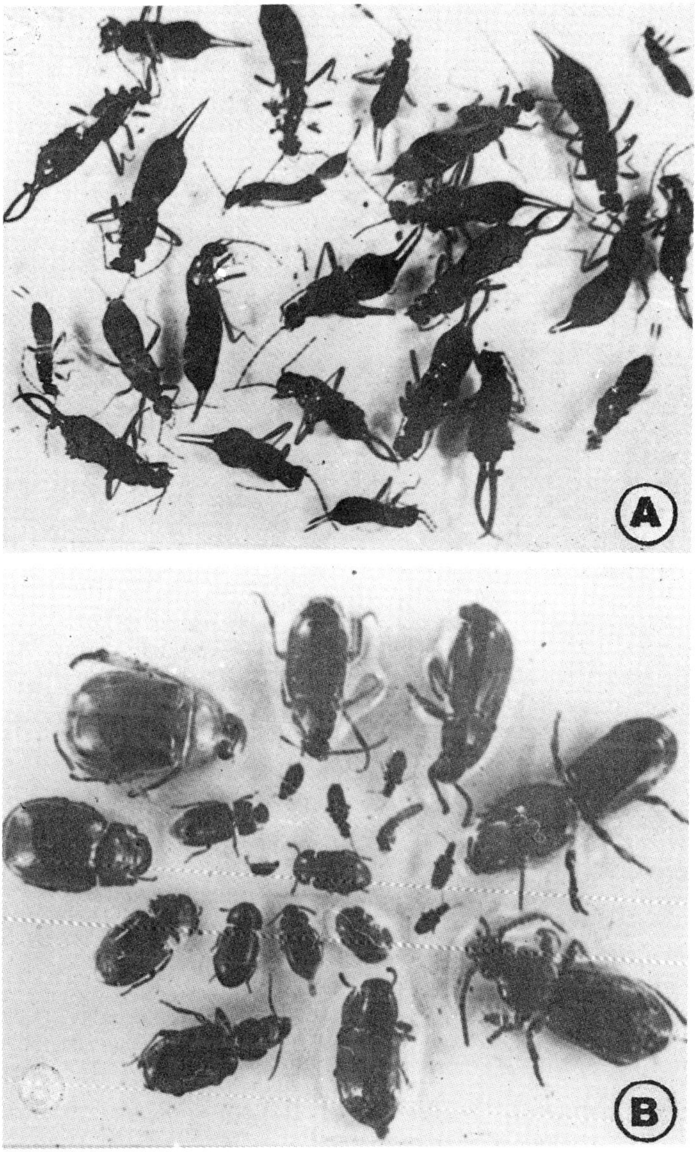

Fig. 20 Some predaceous Dermaptera (A) and Coleoptera (B) in the litter.

their prey. *Indobisium granulatum* shows vertical migration during March, with a downward trend in June–July and an upward trend in July–November during rainfall. In view of their comparatively low number

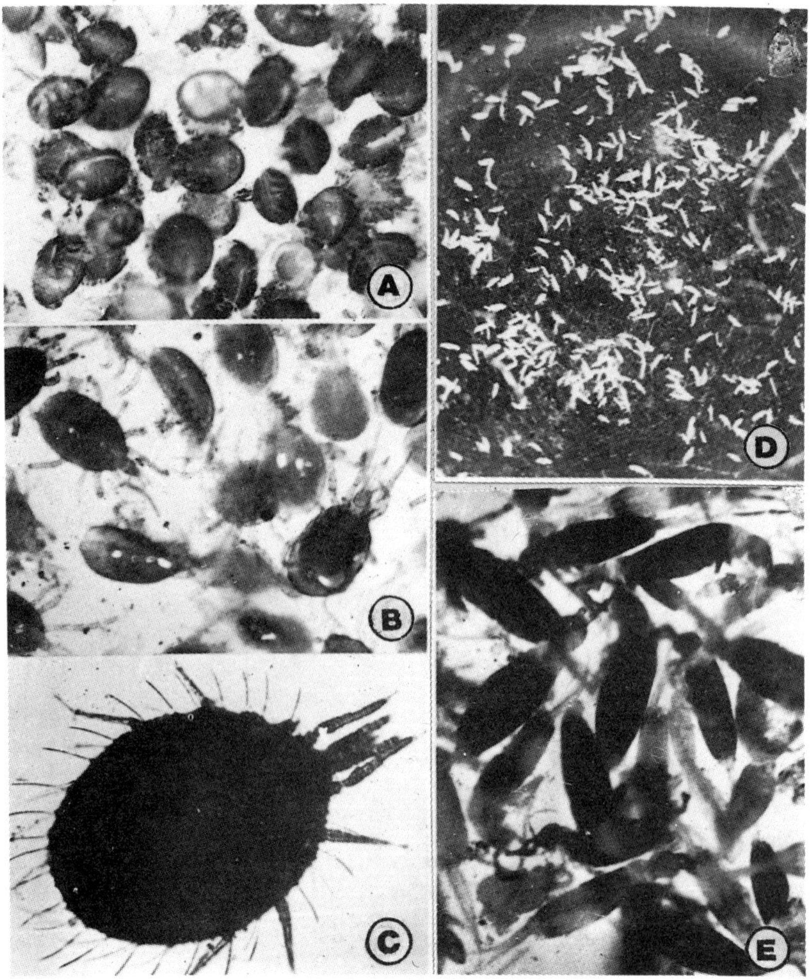

Fig. 21 Predatory mites (A, B, C) and predatory Collembola (D, E).

they may be regarded as tertiary consumers occupying the terminal stage of the detritivore/predatory food chain. However, while feeding on collembolan it may be a secondary consumer (Sabu Thomas, Personal Comm.) (Fig. 22).

While sufficient information on population regulation in natural population of litter microarthropods does not exist, more because of the lack of field-oriented experiments, available data indicates that the size of the population is more regulated by a balance between natality

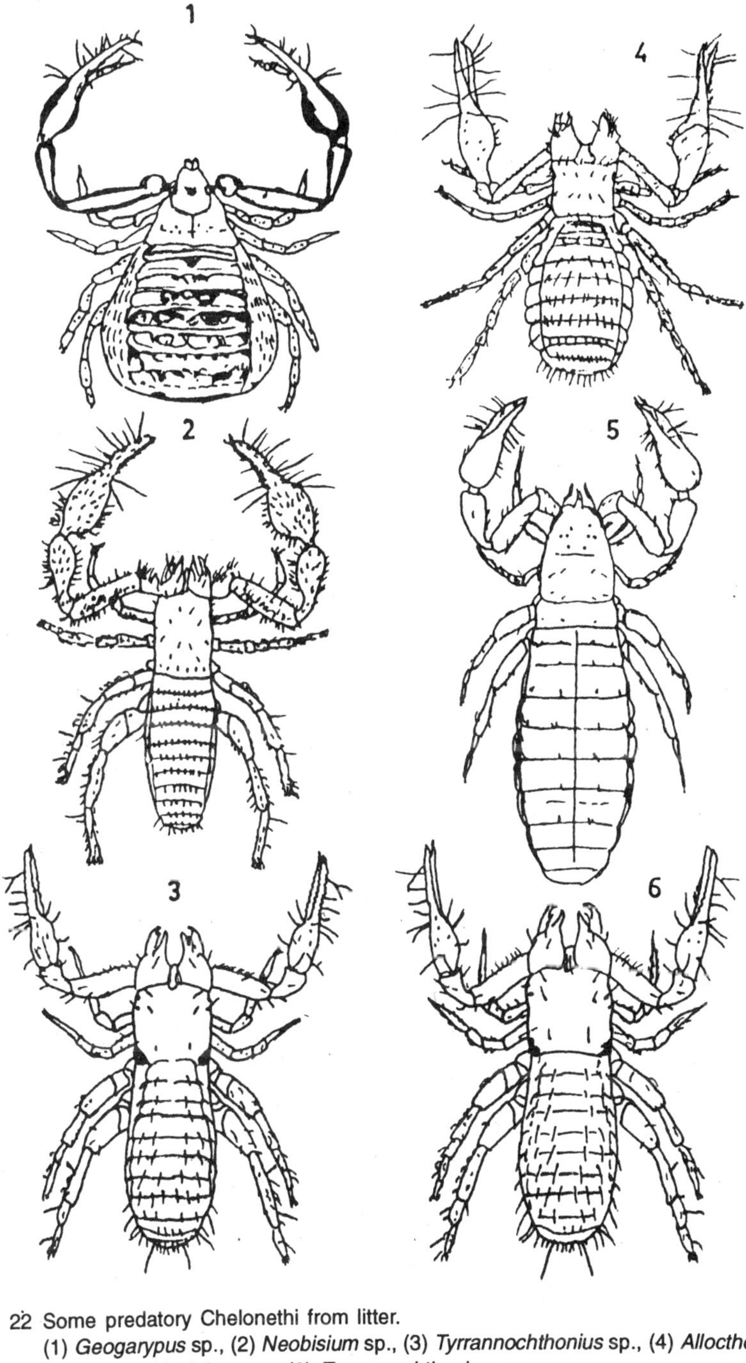

Fig. 22 Some predatory Chelonethi from litter.
(1) *Geogarypus* sp., (2) *Neobisium* sp., (3) *Tyrrannochthonius* sp., (4) *Allocthonius* sp., (5) *Amblyolpium* sp., (6) *Tyrrannochthonius* sp.

and immigration and mortality and emigration (Wallwork 1970), and no mechanism is applicable to litter microarthropods. Predators exert some influence on the population of certain species, and are primary population determinants in the case of Collembola and Acari. While climate and food supply have their own role to play, predation is the main regulating force on primary consumers.

7

Resource Utilisation: An Exercise in Chemical Ecology

Variation in resource quality of litter influences the composition and interrelationships of organisms which are dependent upon it. It is a recognised fact that natural forest ecosystems support more diverse consumer and decomposer systems because of their floristic and accompanying faunistic diversity at various trophic levels, than man-made monoculture vegetation or forests. Reduced structural complexity results from long-term imbalance in interfered forest ecosystems and potentially useful genes are lost when forests are cleared and vegetation cover gets modified leading to a reduction in the stability of the system. The role of forests in stabilising microclimate is well known and the soil cover provided by the litter harbours a wealth of insect resources, both from the viewpoint of the degree of litter availability as well as variability, which of course depend on the litter contributing component of tree species. The preservation and protection of natural polyculture forests to safeguard the natural ecosystem as well as the genetic pool they contain, are important aspects taking into consideration the active role of litter organisms in the breakdown of litter as well as sustaining the food web or trophic complexities in different forest types. Occurrence of size-related systems such as micro-, meso- and macrofaunal elements in their respective trophic levels is again influenced by the resource quality in the respective forest ecosystems. Being the initial food resource on which the trophic structures are developed, the rate of growth of microflora and hence of the grazing fauna indirectly and saprophytic fauna directly are influenced by resource quality. A series of factors determine resource quality, namely composition of carbon and energy providing substances, concentration of essential elements, the presence and concentration of polyphenols which can inhibit or stimulate resource utilization (Hillis 1966: Swift, Heal and Anderson 1979).

It is well known that tropical forest systems have been subject to considerable human interference and undisturbed mixed forests have low incidence of pests as more of the natural regulatory mechanisms are involved. In interfered situations, outbreaks are essentially due to climatic stress, absence of predators and parasites or their ineffectiveness. The biotic resources in litter ecosystems involving diverse insect species and other microarthropods as well as parasites and predatory insects are highly variable in exploited forest areas and natural undisturbed forests as well as in typical monoculture plantations of teak, eucalyptus etc.

Dickinson and Pugh (1974) discuss field variation in microbial and faunal populations associated with different resource types, as well as decomposition of particular types of litter. Resource decomposition, i.e., metabolization of a range of substances with the production of secondary substances, and gradual physical breakdown of resources over time also increase the physically determined niches available to organisms. The impact of chemicals released during litter breakdown and shifts in the relative abundance of primary or secondary compounds including the phenolics, besides C:N and H_2O ratios is another factor that could regulate faunal and microbial diversity and abundance. In this regard cellulose, lignin, hemicellulose, water soluble sugars, amino acids and aliphatic compounds are essentially involved in plant litter. While it is known that earthworms digest a good proportion of cellulose and chitin because their gut enzymes contain cellulase and chitinase, there is a collective degradation by many organisms, so that decomposition of litter and soil faunal activity go hand in hand. As decomposition proceeds, the range of niches expands with the production of metabolites from lignin and cellulose by decomposition and physical disintegration which is particularly associated with faunal colonization. Microbes, collembolans, mites, dermapterans, blattids, carabids, symphylids and diplopods and a host of other insects contribute to the mechanical fragmentation of litter. The quality of litter mass is the primary factor regulating the composition of microbes and fauna in any litter habitat. with the resource quality influencing microbial population it will, in turn, influence the composition of fauna which graze on the microflora.

There is some beneficial effect of soil invertebrates particularly in ecosystems with higher seasonal patterns of litterfall and this benefit results from the fact that their biomass also acts as a reservoir of plant nutrients (Ananthakrishnan 1986). The saprovores consume litter and are, in turn, consumed by predators. Abundance of litter invertebrates, is resource quality-wise superior. Mixed forest litter ensures a wide range of mortality in their respective habitats, so that the nutrients contained in their tissues get mobilized in higher concentrations. This is confirmed by the fact that the tissues of the litter-feeding invertebrates have higher concentration of sodium, phosphorus and nitrogen. Physical and chemical properties of litter contribute to temporal changes in species composition of populations associated with decomposing litter. Certain microarthropods could be used as indicator species for particular stages of decomposition such as a microbial phase, succeeded by an acari phase and finally a humification phase involving collembolans, thrips, oribatids etc.

The quality of fungal food resources for microarthropods may vary in accordance with the mode of nutrition and hence the quality of fungi. A wide variety of fungi are carried by Collembola as soon as they move up and down the organic layer and so have high potential for dispersion of fungal propagules into new resources (Visser *et al.* 1987). However, populations in later stages tend to have large cell size and show faster growth rate than in the exploitation phase.

Interactions between microflora and fauna may include symbiosis in relation to low resource quality. Concentration of nutrients by microflora in nutrient poor resources can gradually provide suitable food resources for fauna, while larger organisms are associated with increase in the quality in litter, and even of the soil. In tropical forests most of the nutrient stock is in the biomass and nutrients in litter shed in the dry season tend to be rapidly mineralised and taken up by roots in the wet season (Dickinson and Pugh 1974). An average of 25 g/m^2/y of litter is assimilated by soil fauna of which 8 g are concentrated into production. The length of the food chain and size of organisms are determined by resource quality which varies considerably between natural forest and man-made monoculture forest ecosystems.

Chemical Ecology in Relation to Forest Litter Dynamics

The vegetative phase of tree growth and development contributes to the release of allelochemicals (Ananthakrishnan 1992), and minerals from foliar surface that are transferred to the nutrient pool in the soil substratum (Molina *et al.* 1991). Litterfall represents an important phase wherein the spontaneous drop off of the senescent foliage and inadvertent fall of young and maturing leaves add to the heterogeneous mix of litter, that functions as a major source of nutrients for the microbial communities and the faunal elements (Usher 1985). The extent of accumulation of subsequent decomposition are intricately related to environmental factors, the impact of which is normally maintained in a state of dynamic equilibrium (Visser 1983 and 1985).

The nature of the allelochemic profiles in the decomposing matrix is essentially a function of the litter source (Vitousek and Sanford 1986) and tends to vary significantly between natural forest systems wherein the litter is contributed by a wide array of plant species and monoculture stands (Ananthakrishnan 1986 and 1988). Uptake, mobilization and mineralisation processes involving nutrient accumulation, utilisation and turnover are biologically mediated processes, and are determined by the functional dynamics of the microbes, the microbial grazers and the whole assemblage of saprophages (Usher *et al.* 1985). The insect community represents a major link in such trophic dynamics by active comminution, channelling and mixing of the decomposing litter, and as microbial grazers alter the inoculation potential and contribute to transfer of spores and dispersal of the microbes (Anderson *et al.* 1981; Swift *et al.* 1979; Visser 1985). Comminution enhances the surface area for microbial colonisation and mediates biochemical transformations. The uptake of nutrients from the litter matrix by the herbivorous micro- and macrofauna further mediates conversions of carbonaceous and nitrogenous substrates contributing to changes in C:N ratio in the litter (Crossley 1977; Seastedt and Crossley 1980; Lee 1991). Insects also serve as useful indicators of changes in the chemical environment (Moursi 1962; Sabelis *et al.* 1984; Bengtsson *et al.* 1988 and 1991; Clement *et al.* 1989) and respond in threshold related modes to semiochemicals emanating from niche resources. With such dimensions of functional diversity contributing to their roles in litter decomposition, an attempt has been made to, (1) correlate the distribution and abundance of insects in terms of the chemical milieu prevalent in the decomposing layers of litter; (2) analysis of structural interrelationships in insect communities has also been attempted to identify differences, if any, between those prevalent in natural forest systems as against the monocultures. Recent evidences (Ananthakrishnan 1988; Annadurai *et al.* 1988; Ananthakrishnan *et al.* 1993) point to a high degree of niche specificity of entomofaunal resources correlatable to the structural integrity of the system and reflect the nature of the biochemical environment. Secondary metabolites including phenolic acids (Katase 1981; Whitehead *et al.* 1981, 1982; Ananthakrishnan *et al.* 1993) and flavonoids have a significant impact on the dynamics of insect communities (Ananthakrishnan and Sabu Thomas 1993), and with progress in decomposition, it is possible that constitution of the coumaric and cinnamic acid derivatives may vary in the L, F, and H layers. Fungi are selectively grazed by the mycophagous insects and several semiochemicals have been implicated (Everaerts *et al.* 1990; Roisin *et al.* 1990; Bengtsson *et al.* 1991) in mediating precise identification of fungal resources even in complex patches. The significance of volatiles released from the decomposing litter *per se* in contributing to the heterogeneity of the chemical milieu and hence in regulating insect community dynamics cannot be overemphasized. An understanding of such dimensions comprising the domain of chemical ecology in relation to insect community dynamics in decomposing litter is central to a holistic evaluation of the interrelationships between the functional groups of insects and their chemical environment. An analysis of such correlations of the functional dynamics of insect communities has been considered in terms of (1) nutrient dynamics in litter, (2) semiochemicals in the litter environment, (3) influence of biochemical variables on litter communities, and (4) biochemical diversity and ecological plasticity.

Nutrient Dynamics in Litter

The biochemically heterogeneous litter matrix is constituted by litter fragments in different states of decomposition, with immobilisation and mineralisation processes in progress, related to the release of nutrients (Coleman *et al.* 1983; Werner and Dindal 1987). With cation exchange status and the levels of organic matter essentially being a function of the structural peculiarities of soil substratum, the intricate relationship between the nature of the forest stand, the phenological patterns of litterfall and hence the nutrient status, and the physiochemcial characteristic of the soil are crucial factors that determine the degree of nutrient release, retention and mobilisation. Such factors as the altitude, temperature, moisture levels and related microbial and insect assemblages have significant roles in mediating rates of C, N, S and P turnover (Tanner 1981; Vitousek and Sanford, 1986). While it is true that the overall productivity in tropical systems is higher than in temperature situations, the dynamics of litter disappearance and decomposition rates vary within the tropical systems to a great extent. The microarthropods represented by collembolans, mites, termites and other litter inhabiting comminutors play significant roles in mediating such transformations (Crossley 1977; Swift *et al.* 1979; Seastedt and Crossley 1980; Ineson *et al.* 1982; Anderson *et al.* 1983; Seastedt 1984). Experiments based on selective colonisation (Reddy 1984; Reddy and Venkataiah 1989) have clearly demonstrated the active participation of microarthropods in the mobilisation of minerals with related changes in the levels of carbon and nitrogen is regulated by abiotic factors. Recent investigations (Ananthakrishnan *et al.* 1983) on the relationships between the levels of utilisable nutrients, viz. major minerals, carbohydrates, proteins and amino acids, and the secondary metabolites represented by the levels of total phenols and the abundance of insect groups have clearly demonstrated an intricate relationship between insect abundance and changes in the levels of metabolites with respect to decomposition. A reducing trend in the levels of metabolites was evident with concomitant increase in the levels of phenols. Significant correlations were also evident between the levels of metabolites and insect population sizes. Collembola dominated the assemblage in the litter of *Tectona grandis*, compared to the preponderance of mites in *Eucalyptus globulus* litter. Dermaptera occurred in significantly high numbers in the latter system. The mobilisation of Na, K, Ca and Mg varied in between monocultures as well as between different layers in each monoculture litter habitat (Fig. 23).

The litter of natural forests derived from about 40 tree species in the moist deciduous forests of the Western Ghats harboured a greater diversity of insects; however, the relative abundance did not vary as a linear function. This suggests the prevalence of dominant species of certain insects along with residual populations of other representatives at any one given point of time. The microbial grazers are the major control links in the active, passive and slow transformation cycles (Anderson *et al.* 1983). Polyphenols tend to alter the biochemical suitability of the niches (Foster and Martin 1981; Ananthakrishnan *et al.* 1993). The transformation of phenolic acids and flavonoids was

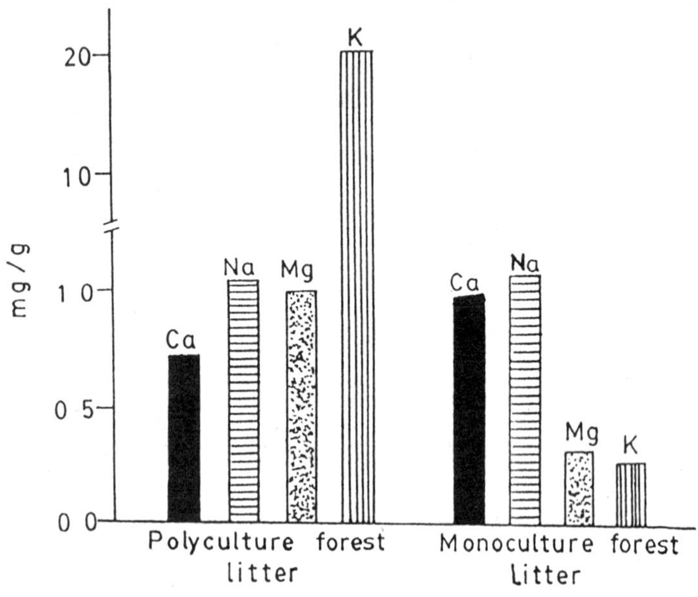

Fig. 23 A comparative assessment of the major minerals in the natural polyculture and monoculture (*Eucalyptus globulus* and *Tectona grandis*) forest-floor litter.

evident in different layers of the litter of teak and eucalyptus plantations and natural forests with several derivatives occurring in different combinations. These included gallic acid, parahydroxybenzoic acid, pyrogallol, phlroglucinol, apigenin, quercetin, kaempferol and chrysoeriol.

Monoterpenes have the potential to alter rates of nutrient cycling, in particular, nitrogen cycling. Extracts of various litters may inhibit nitrification, but persistence of monoterpenes in litter is not well known. Available information on ponderosa pine litter shows total monoterpene concentration was 1,400 μg/g in fresh litter, about 150 μg/g in a year-old litter, 30 μg/g in F-H organic horizon and 0.6 μg/g in 0-10 cm depth mineral soil. Examination of litter for 3 different forest types show that high concentration of monoterpenes in the F-H horizon are negatively correlated with nitrification in F-H and mineral horizon (White 1994). Factors associated with poor litter quality such as high C/N or lignin/nitrogen ratios, which usually show decomposition, may enhance persistence of monoterpenes. Favourable moisture and temperature regimes may enhance microbial breakdown of monoterpenes (White 1994).

Increased level of N_2 and P following excretory releases by the faunal elements in the processes of mineralisation, stimulate decomposition and contribute to ecosystem stability and resilience (Prabhoo 1971; O'Neill 1976; Pimm and Lawton 1978). Successional stages in forest floor tending towards maturity are characterised by high C:N ratios of 40–50:1 (Coleman *et al.* 1983). In mixed deciduous forests, nitrification depends on available NH_4^+ (Vitousek and Sanford 1986; Murphy and Lugo 1986) released in the process of biologically mediated transformations (Figs. 24, 25)

Termites consume up to 17–90% of woody litter and in the process of foraging, enhance the spatial heterogeneity of organic matter (Buxton 1981). The decomposition litter of dry deciduous forests in North India are characterised by high levels of nitrogen (1.2–2.6%), non-structural carbohydrates (2–7%) and water soluble carbohydrates (16–20%), wherein Collembola gain preponderance over the other groups of insects. The rates of decomposition vary from 1.6 to 2.7 mg/day as compared to from 1.07 to 13.1 mg/day in the typical tropical system (Singh 1969).

Semiochemicals in the Litter Environment

Decomposing litter releases several volatiles with are also specific to each forest habitat and also specific to different layers in the same forest habitat itself, as evidenced in the case of *Tectona grandis* and

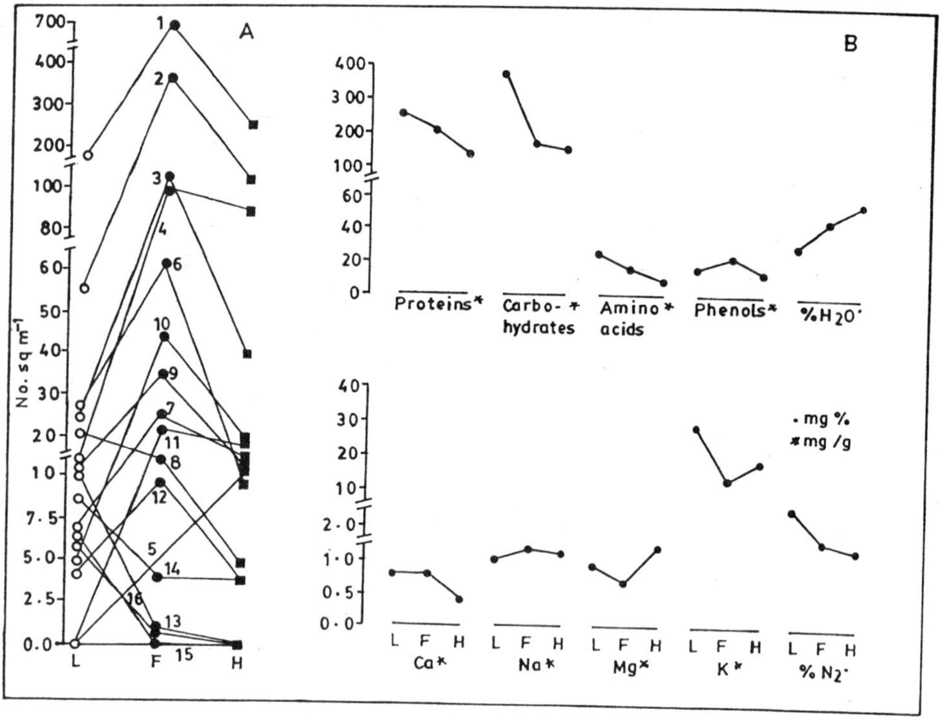

Fig. 24 Populations (no. m^{-2}) of different groups of litter insects (A) and levels of primary and secondary metabolites and in the L, F and H layers of polyculture forest-floor litter at Idikki (B).
1. Mites, 2. Collembola, 3. Coccinellids, 4. Coleoptera, 5. Termites, 6. Thrips, 7. Scarabaeids, 8. Staphylinids, 9. Carabids, 10. Dermaptera, 11. Chelonethi, 12. Blattids, 13. Reduviids, 14. Gryllids, 15. Ants, 16. Dipterans.

Eucalyptus globulus (Ananthakrishnan *et al.* 1993) and natural forest-floor litter. such compounds as Decane, 2,3,5-trimethyl and Decane, 2,3,6-trimethyl were detected in all the three layers. The array of volatiles identified from various forest floors are also indicated (Table 15).

Several volatiles are implicated in influencing the trophic relationships of Collembola and the community structuring of termites. Nutritional resource quality has a significant impact on the reproductive success and related '*r*', '*k*' and '*a*' selection mechanisms and the mycophages respond in specific modes to volatiles released from favourable fungi. Mites exhibit a hierarchical preference for fungal food source and exhibit relatively earlier attainment of reproductive maturity and greater survival. The mycophagous Collembola clearly differentiate between suitable fungi and the unpalatable ones by responding to

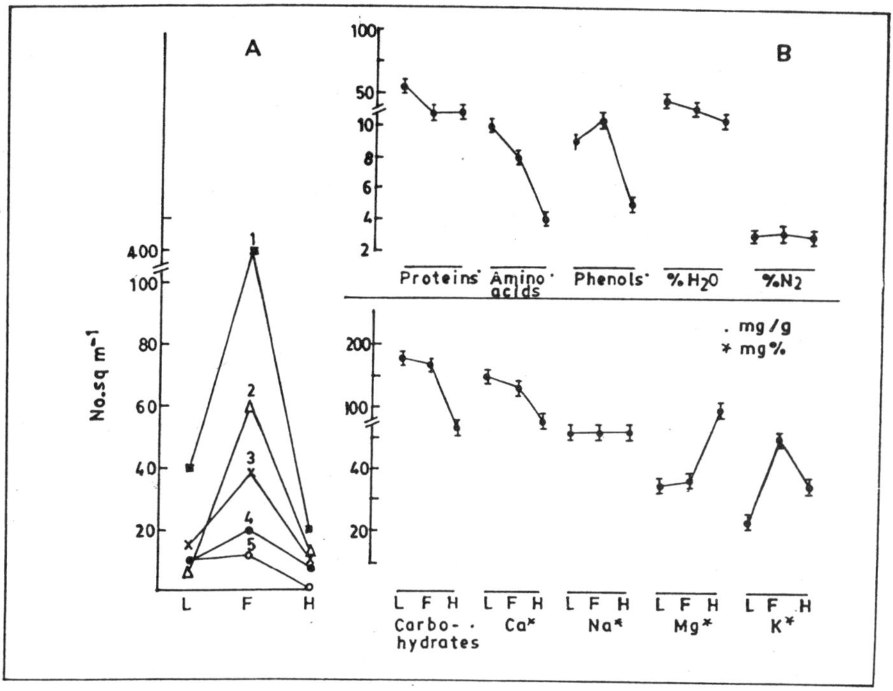

Fig. 25 Populations (no. m^{-2}) of different groups of litter insects (A) and levels of primary and secondary metabolites and minerals in the L, F and H layers of *Tectona grandis* litter at Walayar (225 m).
1. Collembolans, 2. Termites, 3. Dermaptera, 4. Mites, 5. Scarabaeids.

semiochemicals emanating from various colonies (Leonard 1984; Shaw 1988; Bengtsson *et al.* 1988) in the niche. The constitution of odours depends on the chemical nature of the substrate on which the fungi grow and varies with the age of the colony. Such volatiles may either occur as single compound release, characteristic of the fungus, or may elicit responses from Collembola, even as a mixture of compounds in specific ratios released from several colonies (Visser 1986). Experimental evidence points to an intimate relationship between the constitution of volatiles released by fungi grown on either soil extracts or on agar media, as eliciting different responses, even from the same collembolan, *Onychiurus armatus*. Such products of microbial activity as CO_2, ethanol, 2-methyl-1-propanol are produced along common metabolic pathways and their release into the litter matrix elicits specific responses from the fungivores. Collembola locate fungal colonies by perceiving CO_2 released (Moursi 1962). This can be equated to the ability of herbivorous insects to locate the plant hosts on perceiving the alcohols and aldehydes released in plant odours (Visser 1986). Species and habitat-specific volatiles released from fungal colonies aid Collembola in locating a palatable fungus even within a complex patch of high levels of microbial activity. For instance, 1-heptene, and 1-octen-3-ol elicit positive responses from *Onychiurus armatus,* when *Verticillium bulbillosum* is cultured in agar, while the same collembolan prefers *Mortierella isabellina* which releases camphor and 1-methyl-phenylethanone, when cultured on soil isolates (Figs. 26, 27) Compounds such as 3-methyl-1-butanol and v-octalone are also commonly encountered in soils on being released by several species (Bengtsson *et al.* 1991). Decanal and 2-octene arrest Collembola in microcosms, while 1-pentanol and 1-heptene elicit a stop for searching and trigger feeding on acceptable fungus (Bengtsson *et al.* 1988). Such modes of interactions mediated by an array of volatiles are also observed in the predatory behaviour of mites (Sabelis *et al.* 1984). Behavioural responses to niche resources are also a function of threshold saturation, lower levels of which aid in detection of even minute levels of elicitors. Collembola exhibit responses to even nannogram quantities of volatiles related to very low thresholds. C_6 to C_9 structurally simple volatiles are emitted by aerial plant tissues (Bradow and Connick 1990), many of which are alcohols, aldehydes and ketones (French 1985), with significant insect semiochemical activity also with a bearing on fungal growth wherein such factors as carbon chain length and biological activity are related (Bradow 1991).

Termites fragment litter and mix the organic substrates with the litter matrix in the process. Different groups consume woody tissues and humus derivatives. Symbiotic flagellates also aid in cellulose decomposition. The workers of the species *Reticulitermes virginicus* were found to release (3z, 6z, 8z)-3, 6, 8-dodecatrien-1-ol subsequent to feeding on litter colonised by *Gloephyllum trateum.* This compound functions as a trail pheromone which is detected at very short distances. Other compounds include diterpenes which reduce the rate of evaporation of monoterpenes and so maintain a suitable gradient of concentration for longer periods of time. Limonene and myrcene enhance the activity of α-pinene synergestically (Clement *et al.* 1989).

Monoterpenes have the potential to alter rates of nutrient cycling in particular nitrogen. They are potential organic inhibitors. Extraction of various litters may inhibit nitrification, but persistence of monoterpenes in litter is not well known. Available information on ponderosa pine litter shows that total monoterpene concentrations were 1,400 μg/g in fresh litter, about 15 μg/g in a year-old litter, 30 μg/g in the F-H organic horizon and 0.6 μg/g in 0-10 cm depth mineral soil. Examination of

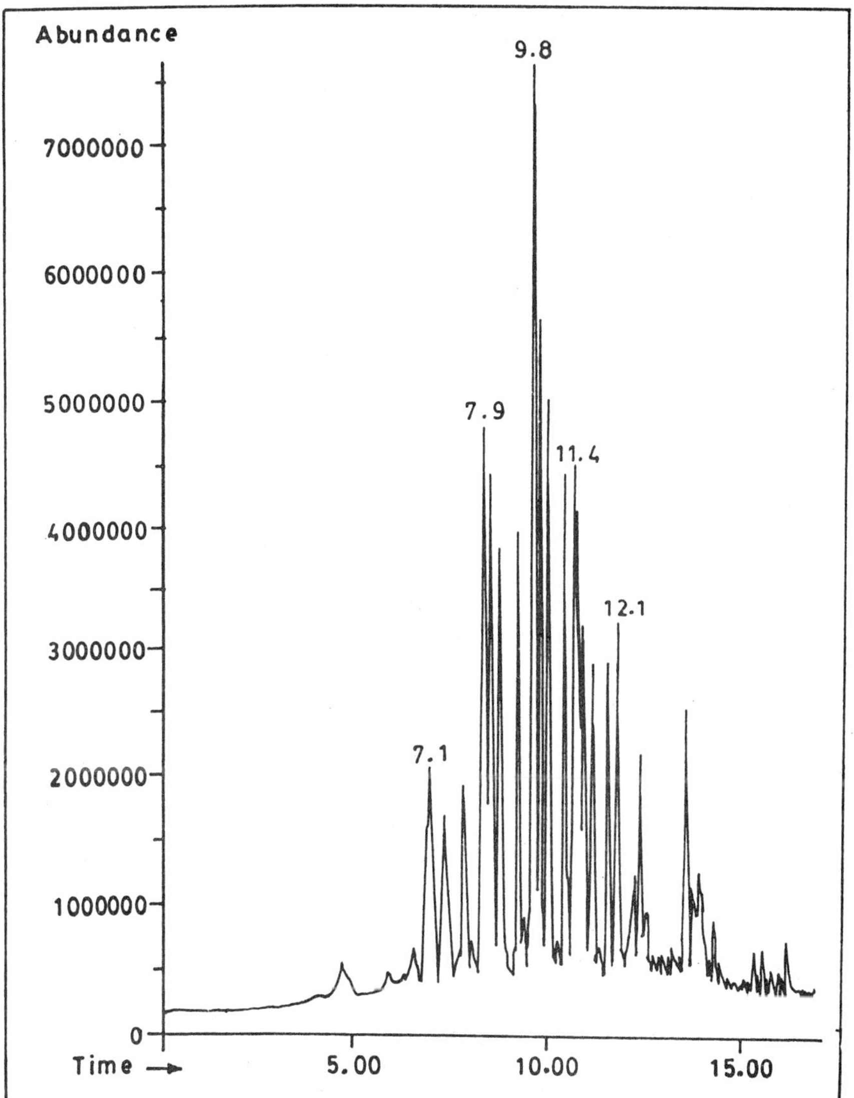

Fig. 26 Gas chromatograms of the volatiles identified in the litter of *Eucalyptus globulus*.

litter from three different forest types show high concentration of monoterpenes in the F-H horizon. Factors associated with poor litter quality such as high C/N or lignin/N ratio which usually slow decomposition, may enhance persistence of monoterpenes. Favourable abiotic conditions like temperature and moisture may enhance microbial breakdown of monoterpenes (White 1994), high C/N or lignin/N ratio which usually slow decomposition, may enhance persistence of

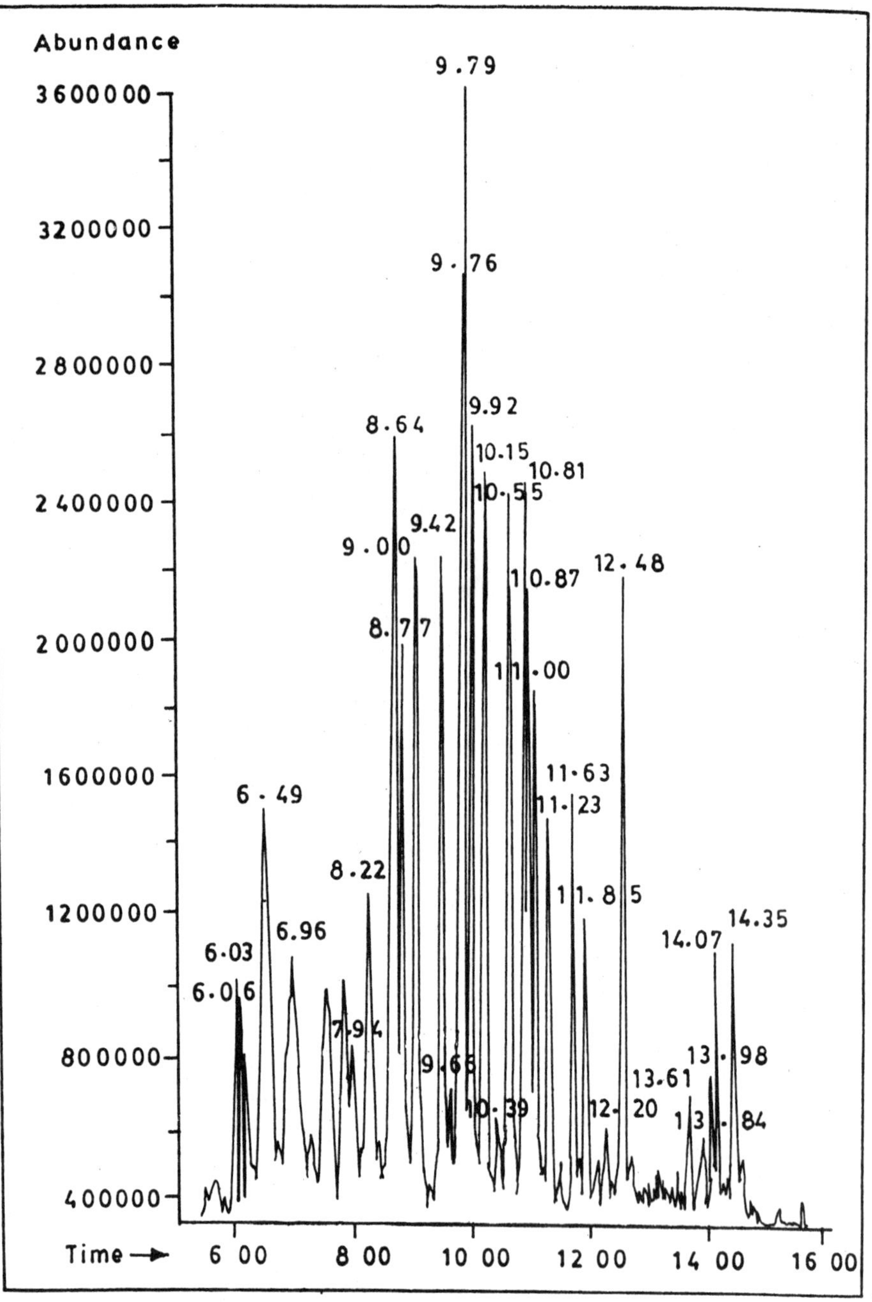

Fig. 27 GC/MS profile of major volatile fractions in freshly fallen litter at Idikki.

monoterpenes. Favourable abiotic conditions like temperature and moisture may enhance microbial breakdown of monoterpenes (White 1994).

Influence of Biochemical Variables on Litter Communities

The role of litter-inhabiting arthropods in comminution and physical breakdown of litter cannot be overemphasized (Butcher *et al.* 1971; Crossley 1977; Seastedt 1984) and the impact of environmental factors in determining their abundance and dynamics is well documented (Anderson *et al.* 1983; Reddy 1984; Wallwork 1959; Witkamp and Drift 1961; Seastedt and Crossley 1980; Swift *et al.* 1979). However, correlations of the dynamics of arthropods with the levels of available primary nutrients, essentially derived from litter, will provide useful indicators of the biochemical changes in decomposition, also in relation to breakdown products including the phenolic acids (Coulson *et al.* 1960). Elevated levels of phenols deter herbivory in the context of which it is important to recognise the relationship between increasing levels of phenols in respect of litter decomposing and their accumulation in terms of the abundance and activity of the insect community found in association with litter. Such an accumulation may shift the balance away from continued feeding on litter.

Collembola and mites prefer to feed on decomposing litter fragments and exploit niche resources not easily accessible to macroarthropods and choose nitrogen-rich fragments. The microphytophagous mites such as *Galumna* and *Ceratoppia* occur also on fresh fallen litter, while certain cryptostigmatids prefer dry litter. The collembolans and mites with their chelate jaws crush the litter fragments between the mandibular plates while some others possess styliform piercing and sucking mouthparts. Most astigmatids and cryptostigmatids are phytophagous and the oribatids have been further classified as either micro- or macrophytophagous forms, while the panphytophagous mites are generalists without any specialised food preferences. The phthiricarids prefer moist litter in a partly decomposed state. The termites feed on woody debris, and the Coleoptera are also saprophagous occurring in all the layers of litter. Microbial grazing consequently results in the release of nutrients into the litter matrix. Nutrient flux dynamics including mobilisation of nitrogen and carbon in particular (Aber 1992; Hobbie 1992) are the most important factors in relation to the nutrient status of the litter. In general, nutrient and mineral releases exhibit an intricate relationship with the saprophagous activity of the mesofauna (Lee 1991; Seastedt 1984).

The chemical environment of the microniches in litter-soil interface is also significantly modulated by the prevalence of phenolic acids and flavonoids (Shindo *et al.* 1978; Katase 1981; Whitehead *et al.* 1981 and 1982) that influence the trophic dynamics of insects (Ananthakrishnan *et al.* 1993). The complex array of volatiles as allelochemicals (Einhelling 1987) further regulate the feeding preferences and mediate location of niche resources. Correlation of the dynamics of insects with the levels of primary nutrients and secondary compounds provide useful indicators of biochemical changes in decomposition.

Comminution of litter fragments and microbial grazing synergise with the saprophytic activity of the microbes effecting release of nutrients including minerals and promote biochemical transformations. Nutrient cycling is also regulated by such factors as species composition, successional patterns and biochemical suitability of the matrix with a bearing on the decay rates. Spatio-temporal biochemical and biological heterogeneity of litter as a function of altitude, and forest cover components further accentuate the uniqueness of each forest litter type. In this context the insect communities in litter are also diverse with several groups carrying out specific functions in the maintenance of structural and biochemical integrity of the complex litter system.

The population density of the fauna may also vary in direct relation to the amount of litter, the age and depth profile in view of the orderly decay sequence and activity of the microflora and faunal assemblages. Yet another factor that could regulate faunal diversity is the impact of chemicals released during litter breakdown and shifts in the relative abundance of primary and secondary compounds, including the phenolics, besides C/N and H_2O ratios. Diversity can also be related to the morphology of litter breakdown indicative of the nature of faunal community, dominant at a given point of time. A correlation of the diversity and abundance of entomofauna with the chemical environment prevalent in L, F and H layers of decomposing litter in monoculture and polyculture stands show considerable diversity.

Analysis of the relative abundance of the 5 groups of microarthropods in *Eucalyptus globulus* litter point to the preponderance of mites in the forest litter layers, and their tendency to increase in very large numbers (2- to 6-folds) in the F-H layers. Collembola tend to persist in relatively high numbers in the humus-soil interfaces. *Eucalyptus globulus* was found to contain relatively high levels of nitrogen (0.8–2% dry weight), reduction in the levels of which could easily be correlated with increase in insect abundance particularly with reference to changes from the L to F litter state. The tendency of mites and Collembola in general to gain preponderance in population densities over the other microartropods suggest their ability to readily exploit a wider range of richness than the other groups. They persist in relatively high numbers even in the humus and can tolerate wider fluctuations in the amplitudes of environmental and biochemical variables.

Dermaptera such as *Euborellia* and *Eparchus* occurred next in abundance and increased 5- to 20-folds in numbers and contributed to nearly 30% of the total assemblage at the highest elevations too. They actively feed on decomposing litter and some are even predatory. Fluctuations in the abundance of mites synchronised with those of Collembola and Dermaptera and as earlier reports suggest, mites (Witkamp and Drift 1961; Wood 1974) serve as good indicators of nutrient status of litter, playing an important role in mineralisation (Seastedt and Crossley 1980; Reddy 1984; Crossley and Hoglund 1962). Collembola and mites were found to gain preponderance over other groups in the litter of *Eucalyptus* species (Reddy and Venkataiah 1989), the activity of which was directly related to weight loss in litter components by biochemical leaching. The relative abundance of minerals varied between the layers. While calcium was leached in highest quantities in the L layers, potassium tended to be released in relatively high levels in the F layers, with humus containing highest amounts in equal quantities.

Biochemical Diversity and Ecological Plasticity

Resource quality influences species diversity and rate of microfloral growth colonizing litter, which in turn, determines the type and intensity of faunal grazing on microbes. These factors together influence increased comminution of litter layers resulting in the regulation of decomposition rates (Anderson 1975; Crossley 1977a; Swift 1976; Swift *ct al.* 1979; Heal and Dighton 1985). Further, the biochemical diversity of litter varies from one forest type to another attaining wider range of variations in mixed natural forest litter compared to monoculture stands (Wallwork 1983; Seastedt 1984; Ananthakrishnan *et al.* 1993a and b). Different species of the forest-floor fauna are distributed in the way they are for a variety of reasons—feeding specificity in some, microclimatic variations in some others or the constraints imposed by the physical and chemical structural peculiarities of the environment in some others.

The abundance and diversity of the microbial community in freshly fallen litter with comparatively low phenolic levels indicated as to how the shifts in microfloral composition are related to litter phenolic levels, as phenolics reduce the palatability and slow down microbial conditioning of litter by acting as feeding deterrents, toxins, and binding agents

(Appell 1993). Species such as *Emericella, Memnoniella, Curvularia* and *Chaetomia* present exclusively in freshly fallen litter and humus layers with low phenolic content have to be considered as highly phenolic sensitive and those without such strata-preference as phenolic tolerant species capable of inactivating phenolic defenses by producing oxidising enzymes (Szklarz *et al.* 1991; Schafer and Blum 1991; Baldwin *et al.* 1983) while procuring nutrients from the substrate matter. High levels of phenols in the F layer and insect abundance show inverse correlation in monoculture plantations (Ananthakrishnan *et al.* 1993a and b). In natural forests no such clear-cut pattern is noticeable emphasizing that either the fauna are phenol resistant (which is less likely to happen) or the diversity of microhabitats and niche availability in the mixed forests together with substrate diversity provide ecological plasticity that will provide diet switching in the face of unpalatability of specific diet items. Predominant fungal species associated with litter habitat, viz., *Aspergillus* sp., *Fusarium* sp., *Penicillium* sp., and *Chrysosporium* sp., have the capacity to metabolize on a wider variety of substrates (Behan and Hill 1978), whereas species such as *Trichoderma, Cladosporium* and *Penicillium* have the ability to grow rapidly and sporulate profusely.

In contrast to monoculture plantations natural forests could be viewed as a major site of retention and accumulation for K > Na > Mg > Ca in the order. Lower calcium levels especially in the L layer and higher levels in the H layer indicate accumulation of calcium in excess as calcium oxalate in the cell lumens of fungi by the secretion of oxalic acid (Crossley 1977). This subsequently reaches the oribatid mites since many forest oribatids feed on fungi which enhance their ability to concentrate biologically important nutrients. Their fluctuation in different litter layers especially in the L and F layers at both forest sites is understandable where maximum fungal diversity and abundance was recorded. Mobilization of fungal nutrient pools and subsequent immobilization in mycophagous fauna consisting mainly of mites, collembolans, thrips and coleopterans is natural, as some of the nutrients will escape with the released faecal matter and arthropodal exoskeletal remains. This is a key event in forest-floor litter trophic relationships.

Interlinking of trophic links whereby detritivores as well as fungi acting side by side as primary consumers was predominant in the fermentation layer indicative of the preference of microbes as well as detritivores towards semi-decomposed litter. Two major energy transfer channels, one starting with mycophagous primary consumers and the other with detritivorus groups are distinct, wherein the former is well established than the latter emphasizing the strong influence of fungal colonisation of augmenting litter decomposition.

8

Chemical Ecology of Litter in Relation to Altitude

Introduction

Faunal turnover along altitudinal gradients especially in litter habitats in tropical forests tend to be rapid enabling an assessment of changes in biomass and relative abundance of taxa, rather than distribution of species. Patterns of altitudinal variation in litter communities are very evident, both in monoculture forests like teak and eucalyptus as well as in natural forests. it is generally known that there is a decline in the abundance and diversity of most taxa and assemblages at higher elevations and a mid-elevation peak in species richness (Olson 1994). The average depth of the litter and humus layers and organic content of the soil increase with elevation as cooler temperatures and water-logged condition slow rates of decomposition (Jordan 1985). Conversely, desiccating condition for leaf litter animals occur in the lowland habitats due to higher temperature and evaporative rates (Olson 1994).

Seven groups of insects were identified in association with litter of *Tectona grandis.* Mites dominated the assemblage in fresh litter followed by the collembolans, the scarabaeids, Dermaptera and staphylinids. Termites and pseudoscorpions were about equal in numbers. Interestingly, the abundance of all the insect groups increased in the F layers except for the scarabaeids. Maximal increases were recorded in collembolans (8–9-fold). The termites were dominant in the humus layer, and the staphylinids also exhibited an increasing trend in abundance, while all the other five groups were recorded in distinctly lesser numbers that in the L and F layers. The levels of cabohydrates, proteins, amino acids, water content and N_2 were significantly high in the fresh litter. With progress in decomposition, a gradual reduction in the levels of these nutrients was recorded. While carbohydrates were lesser by 19–70%, proteins tended to be in equal levels in the F and H layers, subsequent to a reduction by 50%. Depletion of nitrogen from litter was also obvious. Differences in the levels of amino acids tended to range only from 6 to 22%, suggesting their gradual accumulation in the H layer. Total phenols increased with decomposition by 27–28%. Of the minerals, Na and Mg exhibited a decreasing trend with decomposition while Ca and K were higher in the F layers. Reduction in the levels of primary nutrients can be attributed to active uptake by insect groups which exhibit increased population sizes in the F layers. The progressive accumulation of phenols can be considered as a factor detrimental to feeding by insects which have accordingly been found in only much lesser abundance in the H layer, than in the other two.

The diversity of insect groups and their abundance in *T. grandis* litter at 225 m altitude were different. The soil over which litter accumulated was lateritic, with low water-holding capacity. Humus accumulation tended to be very low on account of being washed away by heavy monsoon rains. Partly decomposed litter was, however, 8–10 cm thick in accumulation. In contrast to the L layer at MSL, collembolans dominated the assemblage occurring at more than twice the number/m^2. The mites were just about one-sixth the number. The Dermaptera were marginally higher and termites were the least in abundance. Unlike at MSL, the scarabaeids also increased in numbers (by 25%) in the layers, as was also the trend exhibited by the other four groups. Collembola increased 8–9-fold, followed by termites that were higher by 33% compared to their abundance at MSL. A sharp fall was recorded in the abundance of all the insect groups in the humus layer.

The pattern of changes in the levels of metabolites was similar to what was observed at MSL. However, the extent of depletion was lesser towards the onset of decomposition at the F layer, followed by further reduction in the levels in humus. It is interesting to note that the initial levels of metabolites in litter were higher at this location than what was recorded at MSL. The residual nutrients at the H layer too were higher, except for amino acids, the depletion of which was observed to a greater extent. It is important to note that the level of nitrogen was higher at the F and H layers compared to fresh litter, in contrast to a gradual loss of nitrogen with decomposition. This points to the heterogeneous nature of shift in nitrogen balance and nitrogen to water ratio in litter.

The levels of Ca and Mg were significantly higher in the F layers. While Ca reduced with decomposition, Mg tended to increase 2–3-fold in the H layers. No difference was noticeable in the levels of Na which, however, was much less in concentration compared to the situation in litter at MSL. An upsurge in the level of K was recorded in the F layer followed by a reduction in humus. It is accordingly evident that the mineral constitution varied and the levels of primary metabolites were marginally higher at higher altitudes. The dominant groups of insect fauna varied, though fluctuations in abundance followed comparable patterns in respect of decomposition.

Mites dominated the insect assemblage in fresh litter at 1,200 m altitude similar to the situation prevalent at MSL. Sharp increases in the abundance of Collembola with transition in the F layer followed similar patterns as at 225 m and MSL, suggesting the propensity of Collembola to easily attain large numbers with greater mobilization of nutrients. The staphylinids which were absent at 225 m, were observed at 1,200 m, and along with Dermaptera they exhibited a trend of decrease in abundance towards the H layer. The scarabaeids unlike in the other two situations, also were in larger numbers in the H layer. The increase in termite abundance was similar to the pattern recorded at MSL. Whereas carbohydrates, proteins and amino acids were significantly higher in the fresh litter at 1,200 m altitude (by 12, 33 and 32% respectively), subsequent changes with decomposition followed similar patterns of reduction. The water content was marginally lower and total phenols tended to aggregate in greater quantities in humus, than at 225 m. While Na and K tended to increase with progress in decomposition.

Thin layer chromatographic analyses of litter for phenolic acids and flavonoids revealed the occurrence of myricetin, gallic acid and hydroxybenzoic acid in fresh litter, gossypetin, dihydroquinone, gallic acid in the F layer and kaempferol, gallic acid and in humus. These constituents were found to be similar in the litter derived from all the three sampling areas. The constitution of volatiles identified in the three layers differed significantly. The nature of microfloral components also varied between the sites and in the L, F and H layers. Species of *Penicillium, Aspergillus* and *Fusarium* were common to all the samples analysed.

Correlational analyses point to relationship between the levels of secondary metabolites in litter and the impact of total phenols on population sizes of the arthropods. Conversely, the fall in the levels of nutrients was found to be directly proportional to the abundance of microarthropods. Increased levels of phenols in humus suggests their accumulation, the prevalence of which can be considered an important factor contributing to a drop in the abundance of insects. The termites, however, exhibited a capacity to not only tolerate the relatively higher levels of phenols, but have also increased comparable patterns of rise and fall in abundance in the L, F and H layers at all the three sites. The termites and Dermaptera were in relatively less numbers in the humus at 1,200 m altitude, and increased in numbers with progress of decomposition at MSL and 225 m. The staphylinids and scarabaeids fluctuated in contrasting patterns at MSL and at higher altitudes, whereas pseudoscorpions were evident only at MSL.

Collembola and mites occurred as the dominant faunal elements in the L and F layers. Their abundance, as in the case of other groups followed similar trends of fluctuations. It can be inferred that termites have a relatively high tolerance to increased phenols as is evident from their numbers documented from the H layer. The overall patterns of changes in abundance did not vary with the altitudes, but coincided with the levels of metabolites and the constitution of the phenolic acids and flavonoids. Succession in collembolan communities as in the case of the other groups is also related to the temporal availability of nutrients. They contribute significantly to changes in N_2 ratios (Guru and Panda 1991; Lee 1991) and promote mineralisation. Forest systems irrespective of altitudes always harbour a larger diversity of Collembola and mites (Mukherji and Singh 1970). That they gain preponderance over other groups even in terms of density, has been amply demonstrated by the present investigation. Mesofaunal assemblages in the litter of deciduous forests of north-eastern India (Mukherji and Singh 1970) are dominated by mites, which can also tolerate wide amplitude in the fluctuation of environmental variables. The arthropods function as grazing fauna alongside the saprophagic fungi and accelerate the utilisation of low molecular weight carbon sources, viz., the amino acids and cabohydrates. The fungi associated with the litter layers not only contribute to the observed depletion of nutrients from litter, they also regulate the abundance of mycophages in the density-dependent fashion.

Elevated levels of phenols in the humus layers, with a direct inverse correlation with insect abundance, coincides with a reduction in the levels of primary nutrients and shifts in H_2O/N_2 ratios. The accumulation of phenols tends to be higher in humus associated with clay-loam soil at MSL, compared to a reduction in the humus at 225 and 1,200 m altitudes. Lesser accumulation could be a result of the slower rates of decomposition at higher altitudes attributed to lesser environmental temperatures. The shikimate derivatives have differential impact on feeding by insects (Ananthakrishnan *et al.* 1992). Of the phenolic acids, gallic acid is reported to be most toxic, and has been identified as the predominant component in humus, along with dihydroquinone. These interconvertible phenolic acids and flavonoids (Harborne 1988; Haslam 1985) act as feeding or ovipositional attractants or both governing the suitability of the microchemical environment of insects, the response to which is species-specific. Further investigations on the impact of the individual phenolic acids and flavonoids on the feeding dynamics of litter arthropods will provide a better picture of this dimension in regulating the trophic diversity of the insects observed.

Changes in the levels of minerals also varied significantly between the sites and did not follow any specific pattern. while Ca was detected in marginally higher quantities in the F layers, with similar trends registered in leguminous litter (Durham 1985), a subsequent fall was registered in the H layers at all the sites. The humus was relatively rich in Na, whereas Mg tended to accumulate in the H layers at MSL. It may be mentioned that K being very mobile is less bound to nutrients and hence microarthropods have little impact on K depletion (Coleman *et al.* 1983). The volatiles identified in the latter may have important roles in determining interspecific interactions indirectly affecting their abundance (Ananthakrishnan 1986). The implications of such volatiles in promoting the action of predatory complexes need further investigation.

The pattern of changes in the total utilisable metabolites along different elevations was uniform with a depletion from L to H layer with progress in decomposition with lowest levels at the humus layer. The extent and rate of depletion differed from low elevation to high elevation as carbohydrates, proteins and amino acids were lesser by 43%, 64% and 67% from L to F at low elevations and 24%, 46% and 58% at mid-elevation and 14%, 38% and 22% at high elevation. Depletion of metabolites from L to F and F to H layers increased in direct proportion with the increase in elevations and hence indicating the lesser rate of faunal activity and decomposition rates. Depletion of nitrogen was highest at mid-elevations in contrast to the trend of gradual decline at other sites signifying the unpredictable shifts in the nitrogen balances in natural ecosystems which in part is related to high activity of nitrogen (Figs. 24, 25).

Sodium and potassium levels were higher than that of magnesium and calcium at all altitudes and levels of magnesium steadily rose from L to H layers, whereas that of calcium presented an oscillation with a dip in the F layer succeeded by hike in its levels in the H layer irrespective of the variation in elevations. Sodium and potassium presented an entirely contrasting fluctuation patterns with a steady decline from L to H layers with progress in decomposition. Irrespective of the fluctuation pattern of individual minerals, all the major minerals tended to occur in higher concentration at lower elevation and in lower quantity at high elevations.

Proteins and amino acids showed similar trends in their depletion at the three sites barring the exception that amino acids levels at F and H at lower elevations were more or less same. Phenolic levels gradually rose from L to H layers in direct proportion with the accelerated

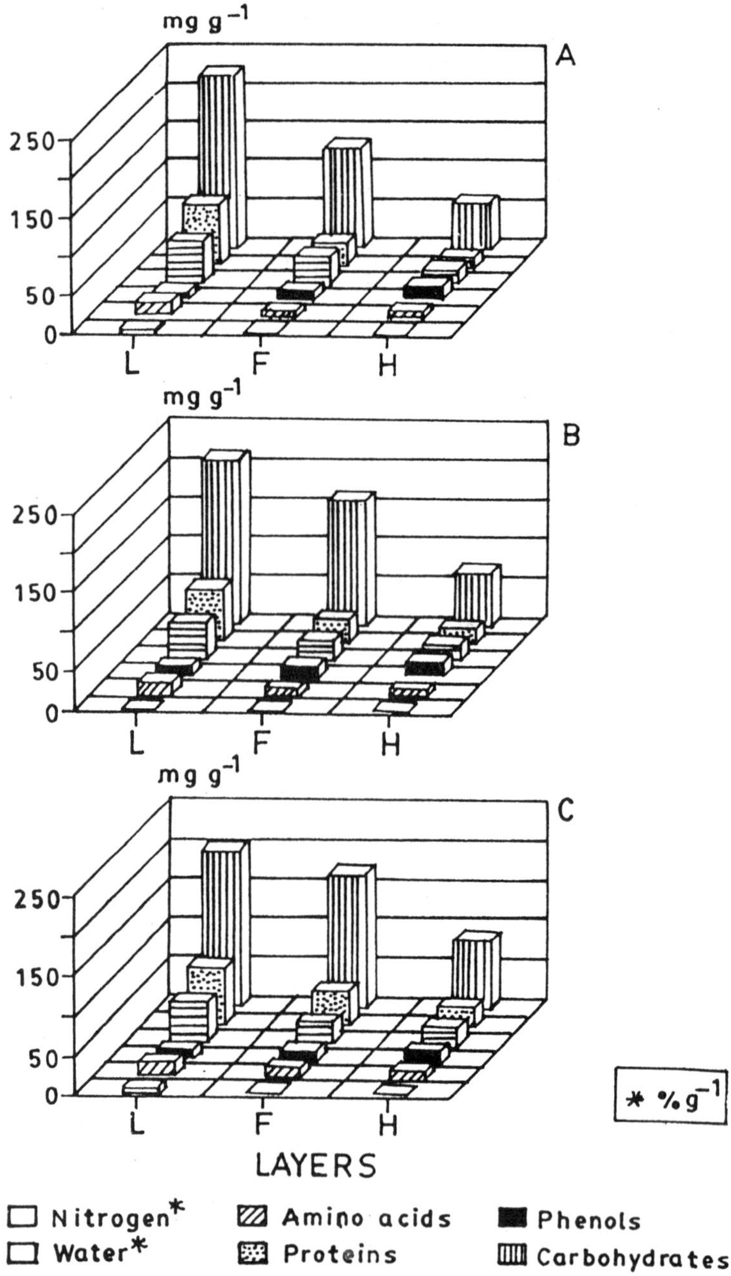

Fig. 28 Levels of primary and secondary metabolites in the L, F and H litter layers along varying elevations (A—1440 m, B—1820 m and C—2220 m) in the Shola forests at Kodaikanal Hills.

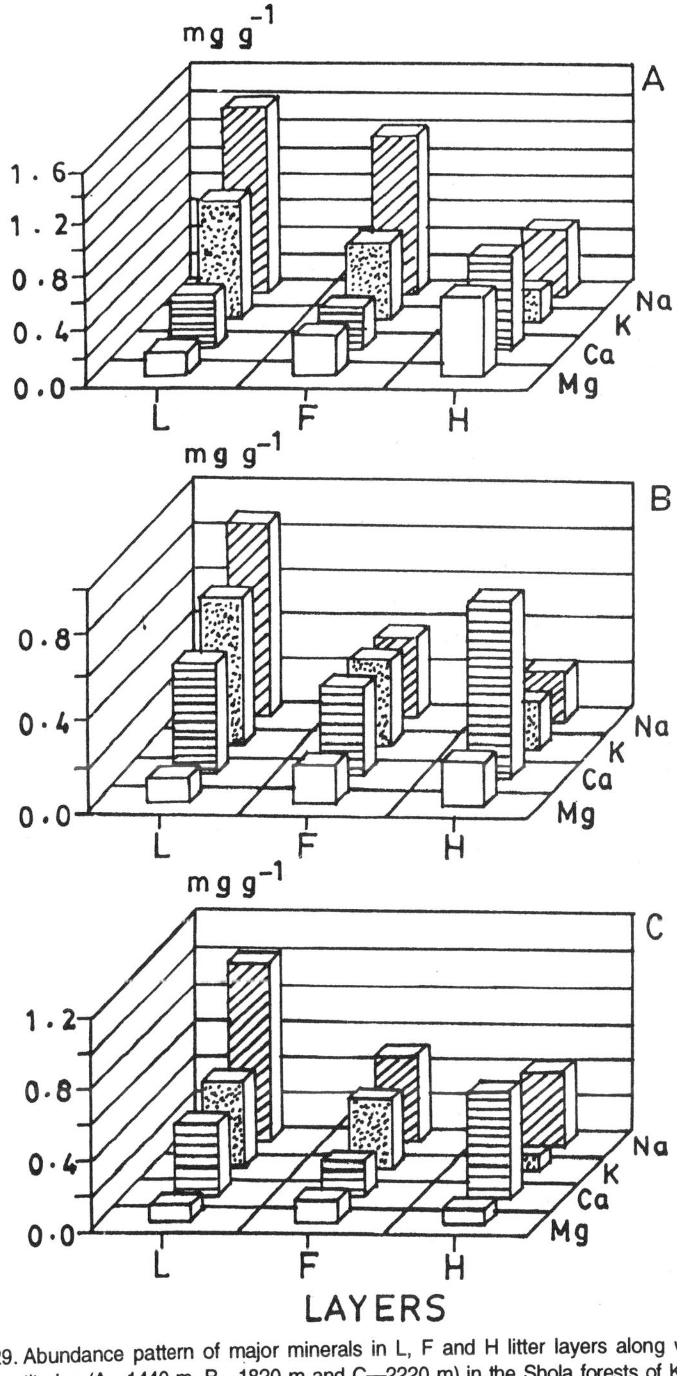

Fig. 29. Abundance pattern of major minerals in L, F and H litter layers along varying altitudes (A—1440 m, B—1820 m and C—2220 m) in the Shola forests of Kodai Hills.

phase of decomposition except at higher elevations. Of particular interest are the levels of phenolics in the mid-elevation humus layer. It is evident that moisture levels peaked in the L layer with maximum values at low elevation, with gradual decline in F and H layers. Salicylic acid, Umbelliferone and dihydroxy quinone was prevalent in almost all the layers at the three sites, mixed with other derivatives. A wide diversity of phenolic acids was detected at 1,440 m and 1,820 m altitudes, whereas the diversity of flavonoids was similar in L, F and H litter zones at all the elevations in Kodai forests.

Volatile profiles in the litter zones showed the prevalence of wide array of complex compounds in all the layers at lower altitudes which decreased in levels with the increase in altitudes. Abundance and diversity of volatiles were highest in the L layer and lowest in the H layer at all the sites (Table 15 and Fig. 30).

Correlation analyses of the interrelationship between faunal abundance and the phenolic levels indicates that rise in the levels of phenols which was maximum in the humus layers at low and mid-elevations is directly related to the population density of microarthropods. Site-specific estimations reveal that at higher elevations levels of phenols were totally in contrast to what has been recorded at lower elevations, since with the fall of phenols faunal abundance dwindled steadily. The fall in the level of nutrients was found to be directly proportional to the abundance of the fauna.

Polyphagous top predators comprising reduviids, dermapterans and Chelonethi prevalent in polyculture forest litter habitat unlike in monocultures exhibits different channels of predation enabling them to have wider prey choices and environmental tolerances to exploit entirely different energy channels by nourishing on primary and secondary consumers as well as on primary predators. Regulatory control possessed by these groups on primary predators, primary and secondary consumers controlled their population dynamics, thus indirectly regulating the rate of decomposition of litter substrate in the forest floor (Fig. 31).

Species abundance and richness of detritivores and fungivores in their respective trophic levels were sufficient to withstand inter- and intraspecific competition, this is more or less restricted in natural forest litter ecosystem in comparison with monoculture ecosystems as a consequence of the wider choice of nutritional sources, microhabitats and niches available for the well establishment of micro- and macroarthropod species groups in the former. That faunal turnover with elevation will necessarily be rapid in ecosystems where many species are restricted to narrow altitudinal belts has been discussed by Olson (1994), with reference to altitudinal surveys in Panama containing large tracts of primary tropical forests. Altitudes sampled were between 300

Table 14. Major phenolic acids and flavonoids in the L, F and H litter layers along the varying altitudes in the Shola forests at Kodaikanal Hills

Localities	Litter layers	Phenolics	Flavonoids
Perumalmalai (1,440 m)	L	Salicylic acid Umbelliferone p-hydroxy benzoic acid	Apigenin Kaempferol
	F	Umbelliferone 3-4-5-trimethoxy cinnamic acid dihydroxy quinone	Kaempferol
	H	Umbelliferone 3-4-5-trimethoxy cinnamic acid	Apigenin Kaemferol Myricetin
Shenbaganur (1,920 m)	L	Salicylic acid p-Coumaric acid p-hydroxy benzoic acid Umbelliferone	Apigenin
	F	Salicylic acid p-Coumaric acid 2-methyl resorcinol p-hydroxy benzoic acid Dihydroxy quinone	Apigenin
	H	p-Coumaric acid Dihydroxy quinone 2-methyl resorcinol p-hydroxy benzoic acid	Apigenin Gossipetin Kaempferol
Pambar (2,220 m)	L	Salicylic acid Dihydroxy quinone	Apigenin
	F	Salicylic acid Vanillic acid Umbelliferone	Kaempferol
	H	Salicylic acid Umbelliferone Vanillic acid Gallic acid	Kaempferol Apigenin

and 2,020 m. Progressive alteration with elevation with over 50% of the characteristic species being lost after a 500 m change is an interesting fact. Decline in species diversity with elevation that the important determinants of local diversity are the continuously varying physical parameters which change with altitude (Table 16). The graphs indicate the overall abundance of species at different elevations, the total number of particular groups such as Formicidae, Staphylinidae, Pselaphidae and Araenae, besides the number of individuals in different taxa at each elevation (Figs. 32–34).

The rate of utilization and immobilization of calcium, sodium, potassium and magnesium is an order of magnitude higher in natural forest stands where faunal and floral abundance and richness is higher.

Table 15. GC/MS profiles of the major volatiles in the L, F and H layers along varying elevations in the Shola forests litter of Kodaikanal Hills

Sites and elevations					
Perumalmalai (1,400 m)		Shenbaganur (1,820 m)		Pambar (2,220 m)	
Volatiles	Area (%)	Volatiles	Area (%)	Volatiles	Area (%)
L-layer					
Pentadecanoic acid	96	9,12,15-Octadecatrienoic acid	93	2,6,10,15-tetramethyl Heptadecane	96
1-propylnonyl-Benzene	94	4-phenyl Pentadecane	91	Heptacosane	90
1-phenylmethyl-Aziridine	90	2-phenyl Pentadecane	91	2,6,10-trimethyl Tetradecane	89
3-Tetradecen-5-yne	90	2-propyl 1,3,2-Dioxarsenane	90	Tetradecanoic acid	81
1-methyldecyl-Benzene	87	9,12,15-Octadecatrienal	87	2-dicarboximid 4-Cyclohexene-1	78
Undecanoic acid	85	1-propyloctyl Benzene	83	butyl 1,2-Benzenedicarboxylic acid	76
Heptacosane	85	1-ethylnonyl Benzene	83	Demecolcine	64
1-ethylnonyl-Benzene	74	1-butylheptyl Benzene	81	Tricyclo [4.3.1.13, 8] undecane	64
bis 1,2-Benzene dicarboxylic acid	74	Undecane	81		
1-butylnexyl-Benzene	64	6-phenyl Pentadecane	80		
1-butylheptyl-Benzene	64	3-ethenyl Cyclooctene	72		
1-butyloctyl-Benzene	64	butyl 1,2-Benzenedicarboxylic acid	64		
F-layer					
1-2-dimethoxy-4-2-prophenyl Benzene	94	6,10,14-trimethyl-2-Pentadecanone	83	butyl 1,2-Benzenedicarboxylic acid	68
				24-dien-26-oic acid Ergosta-2	64
Hexadecanoic acid	91	2,6-dimethyl-, methyl Nonanonic acid	72	Silver (1+) salt Octanoic acid	64
3,7,11-trimethyl 1,6,10-				6-bromo-Hexanoyl chloride	64
Dodecatrien-3-ol	90	2,8-dimethyl-, methyl		dipropyl 1,2-Benzenedicarboxylic acid	64

Table 15. (*Contd.*)

2-methyl-1,3-Butadiene	85	undecanoic acid	72	diheptyl 1,2-Benzenedicarboxylic acid	64
1-3,8-P-MENTHATRIENE	83	2, 10-dimethyl-, methyl			
1,2,4a,5,8,8a-hexahydro-		Dodecanoic acid	72		
Naphthalene	83	bis 1, 2-Benzenedicarboxylic			
N-1,1-dimethylethyl Benzamide	64	acid	64		
2-bromo-1-4-methoxyphenyl Ethanone	64				
1,2-dimethoxy-4-1-propenyl Benzene	64				
Tetradecanoic acid	64				
H-layer					
butyl 1,2-Benzenedicarboxylic acid	90	3-nitro 1,2-Benzenedicarboxylic acid	96	butyl 1,2-Benzenedicarboxylic acid	86
Pentadecanoic acid	80	2-methyl-Eicosane	93	Pentadecanoic acid	80
1,3; 2,4-di-o-benzylidene-		4-methyl-Eicosane	89	3-methyl-Pentane	64
Erythritol	64	Hexatriacontane	89	dicyclo 1,2-Benzenedicarboxylic acid	64
bis 1,2-Benzenedicarboxylic acid	64	2,6,10-trimethyl-Tetradecane	89		
1-oxide 2-Quinolinol	60	Tricyclo [4.3.1.13,8] undecane-1-carbox	87		
		Pentacosane	83		
		3-methyl-Hexadecane	81		
		3-methyl-Nonadecane	74		
		2-methyl-8-propyl-Dodecane	74		
		5-methyl-2-phenyl-1H-Indole	70		
		Tridecane	68		
		2-methyl-Nonadecane	68		
		2-Phenyl-1-Benzopyrylium	64		

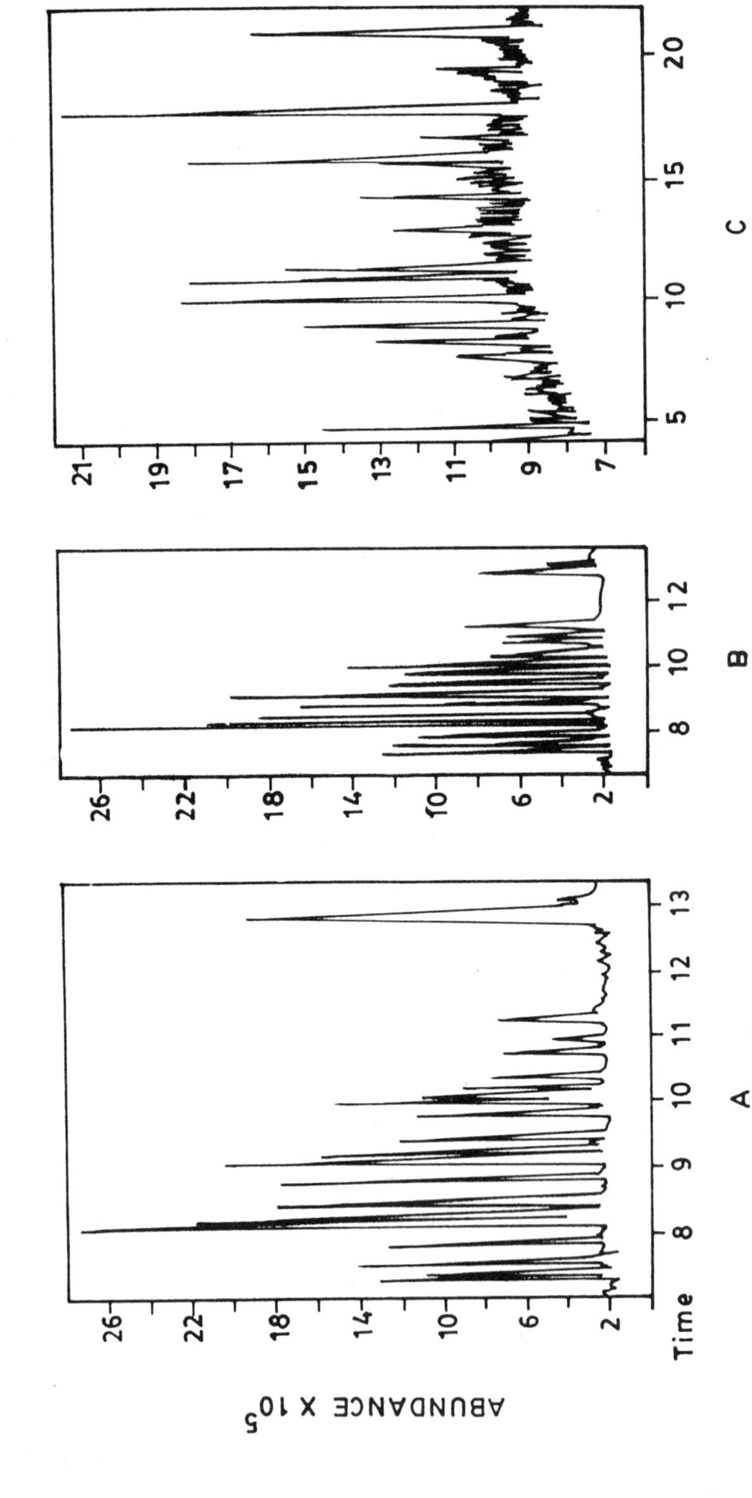

Fig. 30. GC/MS profile of volatile fractions in the L-litter layer along three elevations (A—1,440 m, B—1,820 m and C—2,220 m) in the Shola forests at Kodai Hills.

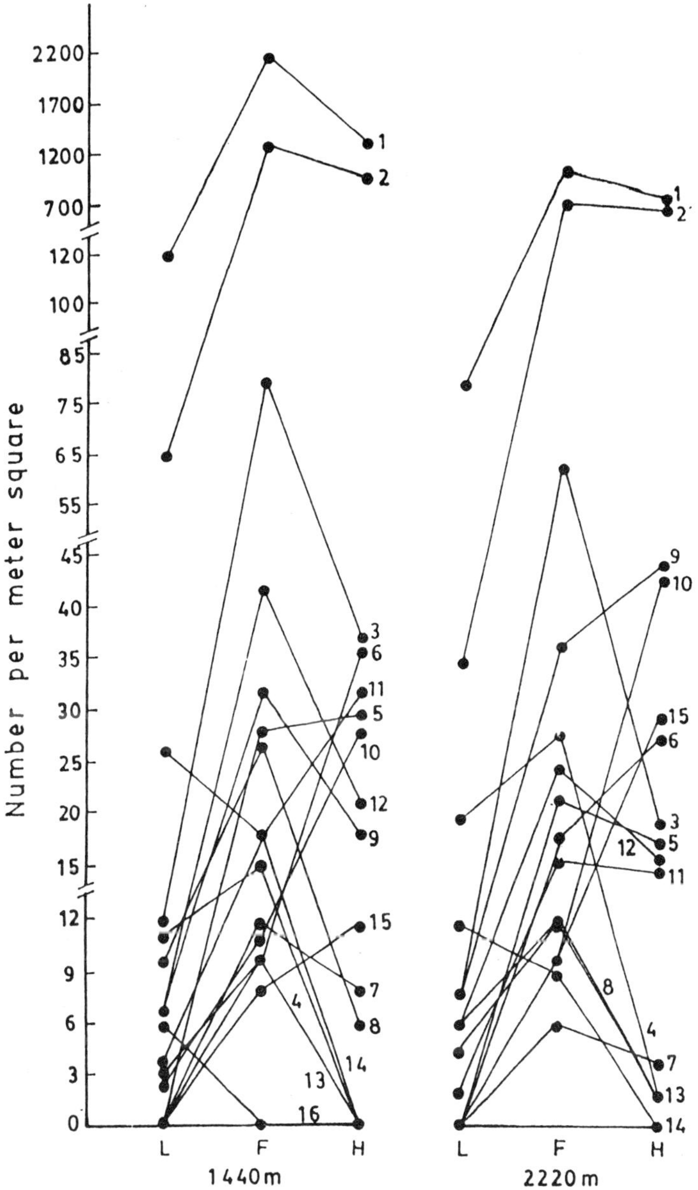

Fig. 31 Populations (no. m^2) of 1. Mites, 2. Collembola, 3. Dermaptera, 4. Ants, 5. Pseudoscorpions, 6. Blattids, 7. Termites, 8. Thrips, 9. Staphylinids, 10. Carabids, 11. Scarabaeids, 12. Other Coleopterans, 13. Gryllids, 14. Dipterans, 15. Grubs—Coleopterans, 16. Reduviids in the lower (1,440 m) and higher (2,220 m) elevations of the Shola forest litter at Kodaikanal Hills.

Table 16. Number of individuals in different taxa at each elevation (After Olson, 1994)

Taxa	Sample elevation (m)							
	300	500	800	1,050	1,250	1,500	1,780	2,020
Formicidae (total)	1614	739	1292	1846*	1362	175	35	33
Formicidae (pitfall)	55	74	113	436	190	74	11	9
Wasps (total)	27	37	116	48	69	15	7	11
Staphylinidae (total)	78	163	406	208	199	40	86	56
Staphylinidae (pitfall)	5	6	13	25	23	5	11	6
Pselaphidae (total)	58	67	174 ■	187	117	17	5	0
Scydmaenidae (total)	39	22	123 ⊙	126	147	25	15	1
Curculionidae (total)	93	34	120 ❖	28	78	114	163	68
Carabidae (total)	5	7	71 ◈	16	12	2	50	143
Carabidae (pitfall)	1	0	2	0	9	0	20	79
Dermaptera (total)	3	9	34	42	8	3	3	1
Araneae (total)	26	18	132	45	64	18	23	4
Araneae (pitfall)	5	8	6	8	13	2	9	0
Pseudoscorpions (total)	41	6	50	12	43	5	2	0
Isopoda (total)	2	4	339 ✪	42	142	6	4	14
Isopoda (pitfall)	2	3	75	11	40	2	3	12
Amphipoda (total) ◊	0◈	4	257	98	188	66	93	17
Amphipoda (pitfall)	0	1	82	5	14	10	12	7
Hemiptera (litter sift)	22	15	48	33	28	15	25	17
Blattodea (total)	24	18	33	32	21	1	42	22
Blattodea (pitfall)	2	2	3	2	3	1	14	11
Diplopoda (litter sift)	15	15	115	47	94	14	30	35
Chilopoda (litter sift)	6	9	lost	16	14	18	7	1
Total	1957	1081	3038	2604	2523	475	520	388

*593 individuals were from one *pheidole* sp. which may be a swarm raider. ■ 61 individuals from 1 species. ⊙ 55 individuals from 1 species. ❖ 55 Individuals from 2 species. ◈ 62 individuals from 1 species. ✪ 175 individuals from 1 species. ◊a single species. ◈ amphipods have been observed as low as 100 m.

Monoculture plantations exhibited an abundance in the order of Na > Ca > Mg > K, whereas that of natural litter is in the order K > Na > Mg > Ca. Differential patterns of mineral turnover can be related to faunal elements at in litter habitats of monoculture and polyculture forest litter habitats indicates that the polyculture system habitats are more productive and complex in which all the above parameters distinctly exceeded those in monoculture forests. These approaches establish that the intensity of litter decomposition in both the forests is variable and these variations could be correlated with the climate, substrate quality of the litter arising from the heterogeneity and resource

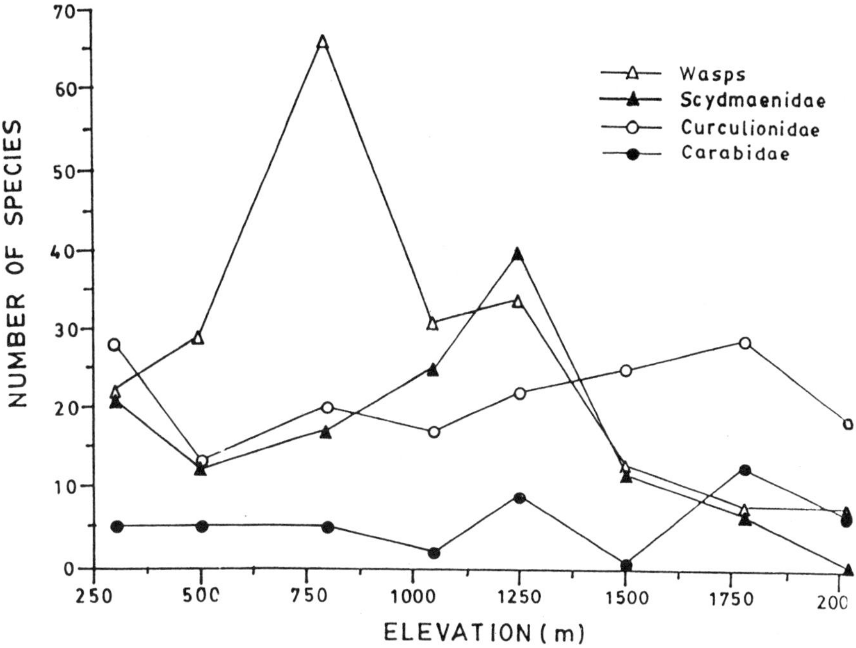

Fig. 32 The number of species of wasps, Scydmaenidae, Curculionidae and Carabidae sampled at each elevation (after Olson 1994).

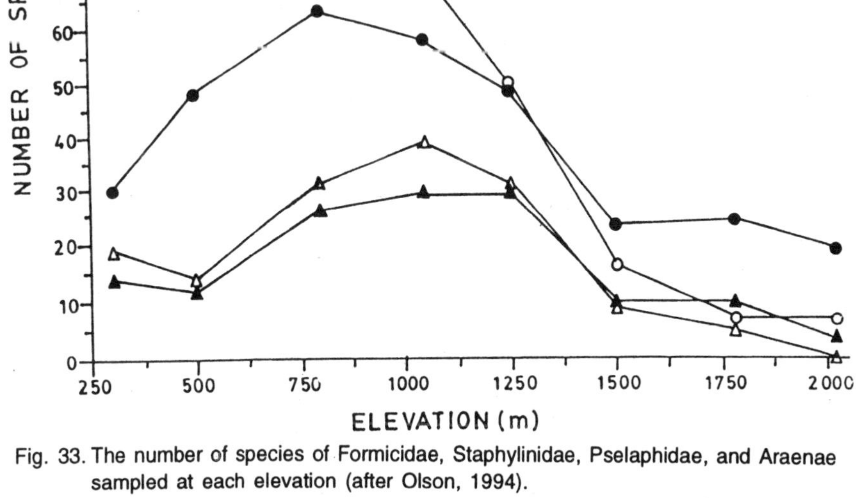

Fig. 33. The number of species of Formicidae, Staphylinidae, Pselaphidae, and Araenae sampled at each elevation (after Olson, 1994).

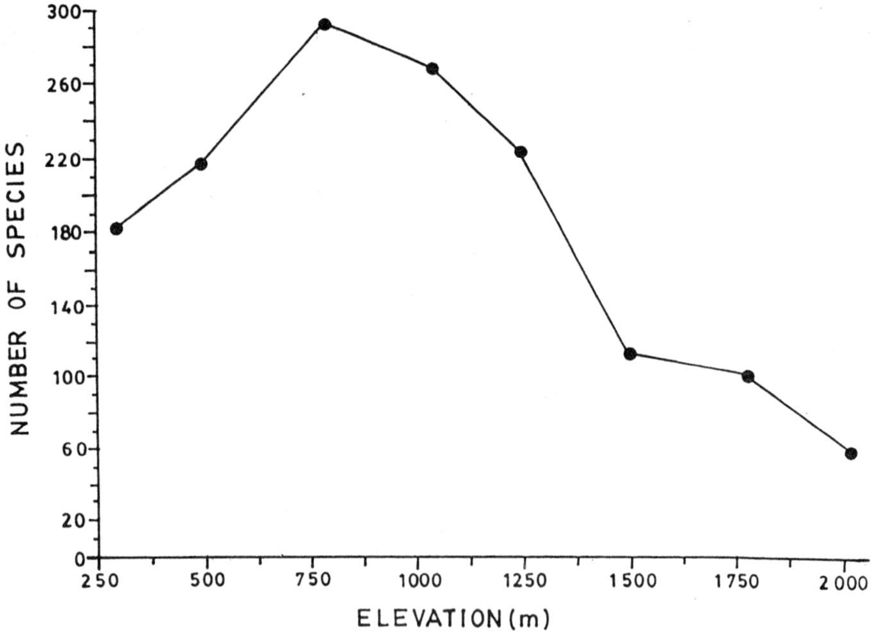

Fig. 34. The total number of species for all tabulated taxa sampled at each elevation (after Olson 1994).

quality of the contributing tree species and also the physical climate prevailing in the habitat. Potential ecological interactions between the microbes and the faunal elements belonging to all the major functional groups and the plant litter, under different levels of disintegration releasing a diverse array of biochemicals establish a complex network in natural polyculture forest-floor litter habitat than in monoculture litter.

9

Trophic Systems in Forest Litter Communities

Introduction

A complex assemblage of species whose number, diversity and ecological interaction play an important role, is typical of forest litter ecosystem. While the trophic position of organisms in the food-web is a dominant factor influencing litter decomposition, trophic levels contribute to the conceptual basis for interpretation of the numerous food-chains present. Litter communities offer good examples for the study of food-webs enabling description of the kinds of organisms in a community which consume or feed on other insects. A community-web, unlike a source or a sink-web, comprises species within a habitat or a set of habitats, without regard to the feeding relation between them, but work out as to who feeds on them (Pimm *et al.* 1990). The analysis of community-web is in terms of trophic species and the shape of the food-web has inevitable dynamic consequences for constituent species. Complex communities should be most sensitive to loss of species from the top of the web, in contrast to simple communities as observed in monoculture forest litter, which are more sensitive to loss of plant species, because the consumers are dependent on one of a few species and cannot survive the loss (Pimm *et al.* 1990). Variation in resource quality is the main basis of trophic structure often influencing the composition and interrelationships of organisms depending on it. The recognition of size-related system such as microtrophic, mesotrophic and macrotrophic also appear useful.

The rate of litter decomposition is influenced by the rates of food consumption and many specific relationships occur in relation to energy utilisation and nutrient metabolism of populations and trophic levels, which are dominant factors influencing these relationships. Trophic levels constitute the basis for interpretation of diverse food-chains present in the ecosystem, the activities of litter species enhancing soil turnover and aeration. Litter fragmentation contributes to microbial decomposition of detritus and eventually the release and recycling of nutrients in the ecosystem (Edwards *et al.* 1970). Food-chains form the major biotic pathways for movement of energy and nutrients, the chemical energy of the food sustaining the biological component. The organisms and the ecosystem have evolved the ability to transform energy, exploit and contribute to the increasing entropy of the system, thereby involving a bioenergetic basis for evaluating system processes (Reichle 1971). Three major energy states are apparent:

(1) Temporary storage of chemical energy in the biomass of the producer and consumer organisms.
(2) Energy losses at each trophic level due to respirating heat exchange.
(3) Energy transferred between trophic level in the feeding process.

Although energy is the driving force, the availability of nutrient elements often affects the rates at which the process occurs. Calorific rate for litter-infesting arthropods has been known to average (for 14 different species) 5,788 cal/g ash-free dry weight (Edwards, Reichle and Crossley 1970). Trophic level efficiencies like intake, association, respiration, growth and production have been shown to be constant for many different species in an ecosystem. The major energy states comprise the temporary storage of chemical energy in the biomass of producer and consumer organisms. Energy losses from each trophic level are due to respirating heat exchange and energy transferred between trophic levels in the feeding process (Table 17). Several trophic levels exist including acarines, Collembola. Araenae, Diplopoda, Pauropoda, etc. The available calorific value for litter organic matter, such as litter (leaves and small twigs) is of the order of

L horizon 4865 (Cal/g ash-free dry weight)

F_1 horizon 4853 (Cal/g ash-free dry weight)

F_2 horizon 4773 (Cal/g ash-free dry weight)

H horizon 4989 (Cal/g ash-free dry weight)

The forest-floor arthropod population is a complex assemblage of species supported primarily by detritus desired for the natural turnover of organic materials. As decomposers, these saprophages play an important role in the breakdown of organic substances. Trophic activities of detritus feeding arthropods and of their predators are intimately involved in the distribution and transfer of nutrients during decomposition process. For trophic level distribution and movement of elements in

Table 17. Whole-body calorific contents of common groups of soil and litter invertebrates

Group	Cal/g ash-free dry wt	Ash percentage
Arthropoda		
Isopoda	6,123	45.2
Diplopoda	5,453	54.9
Chilopoda	5,344	7.2
Phalangida	5,732	6.9
Acarina	5,808	–
Araneae	5,794	3.8
Collembola	6,063	8.7
Orthoptera	5,722	4.0
Coleoptera	5,672	3.2
Diptera (larvae)	5,464	–
Hymenoptera	6,247	2.8

(after Reichle 1971)

food-chains, information on the chemical composition of species is necessary. The data provided by Reichle, Shanks and Crossley (1969) on the whole body chemical composition of Ca, K and Na in forest-floor infestation is given in Table 18.

Table 18. Whole-body chemical composition of calcium, potassium and sodium in forest floor

	Element (mg/g ash-free dry weight)		
Group	Calcium (Ca)	Potassium (K)	Sodium (Na)
Arthropoda			
Diplopoda	103–546	3.2–14.8	1.4–6.0
Chilopoda	0.6	7.1	9.2
Scorpionida	2.3	0.2	7.0
Phalangida	7.2	7.6	5.7
Araneae	2.1	7.2	7.3
Thysanura	4.6	–	4.3
Orthoptera	2.1–4.3	2.4–4.6	5.7–7.1
Hemiptera	0.7–1.8	4.7	1.9–2.3
Coleoptera	0.3–2.5	0.9–6.5	2.7–5.9
Lepidoptera (larva)	4.7	50.0	9.2
Diptera	2.0	1.2	5.4
Hymenoptera	0.7–2.4	1.8–9.4	2.8–3.6

(after Reichle 1971)

Resource Quality and Composition of Faunal/Microbial Elements

Resource quality includes composition of carbon and energy providing substrates, concentration of essential nutrients and polyphenolics which could modify resource quality or which can inhibit resource utilisation. As decomposition proceeds, the range of niches expands with the production of metabolism from lignin and cellulose decomposition, as well as physical disintegration, the latter particularly associated with a phase of faunal colonization. So the quality of resource is the primary factor which selects the composition of the microflora and fauna. With the resource quality influencing microbial populations, it will, in turn, influence composition of the fauna which graze in the microflora, as well as determining directly the composition of the decomposer fauna. The functional groups of the forest-floor faunal complexes at two natural forest sites are indicated in Table 9. Modes of feeding, principal food source, life cycles and habitat essentially reflect functional aspects. Taking the predators for instance, there are the general predators, specialist predators, and omnivores, so that a heirarchial structure can be developed to explain interrelationships of species, as also that among feeding guilds and other functional groups. The importance of these interactions can be assessed through adequate analyses of food webs. In the formulation of a food-web, a study of the interactions among species becomes essential and any community such as the decomposer community has to be studied in its entirety before we understand the dynamics of the system.

There is a sharp decline as readily available substances are exhausted or as fauna exploit the food source. Initial plant resources of high quality ensures energy supply, but with increasing humification resource quality declines leading to competition for the available resources. Increased importance of large organisms is associated with the increase in resource quality, which in turn, determines the length of food-chains and size of component organisms. Further, as the resource decomposes over time, the range of substances is reduced and secondary substances are produced. The gradual physical disruption of the resource over time increases the niches available to organisms.

In general, the population of organisms developing on a resource do not follow a simple pattern. There is often a small-scale mosaic of resources which support association of organisms at varying states of development with distinctive interrelationships.

Trophic Diversity of Some Litter Insects

The species composition in natural forest litter in contrast to that of monoculture, indicates the existence of potential trophic links involving mycetophagous elements such as thrips like *Neurothrips indicus* feeding on *Aspergillus niger* and *Penicillium* sp. (Hyphomycetes); *Stigmothrips limpidus* feeding on *A. niger, Penicillium* sp. (Hyphomycetes) and *Pestalotia* sp. (Coelomycetes); *Ecacanthothrips tibialis* feeding on *A. niger* (Hyphomycetes) and *Fusarium oxysporum* (Hyphomycetes), and *Hoplandrothrips flavipes* feeding on *Aspergillus* sp. (Hyphomycetes). The sporophagous elements include *Dinothrips sumatrensis* feeding on *Verticicladium theobromae* (Coelomycetes); *Elaphrothrips denticollis* feeding on *Pithomyces* sp., *Aspergillus* sp. (Hyphomycetes) and *Pestalotia* sp. (Coelomycetes); and *Nesothrips* sp. feeding on *Aspergillus* sp. (Hyphomycetes). In eucalyptus litter, individuals of *Priesneriana kabandha* were noticed feeding on *Cystospora* sp. and *Rhytidhysterium rufula* (Coelomycetes). Apart from collembolans and acarines, such lygaeids as *Metochus uniguttatus, Graptostethus servus* and *Naphiellus dilutus* served as secondary consumers. The earwig species *Forminolabis sisera, Euborellia* sp., *Eparchus insignis* and *Labidura riparia* and the reduviids as *Pirates affinis, Acanthaspis coprolagus, Lazarda annulosa*, and *Rhinocoris* sp. feed on the other arthropods like ants, carabid beetles, termites and gryllids thereby serving as the top carnivores comprising the predatory food-chain. Anthocorids such as *Cardiastethus* sp., *Scoloposcelis parallelus* and *Orius maxidentex* also contribute to the predator chain (Figs. 35–37)

A comparative assessment of the entomofauna in three different monoculture litter habitats, viz. eucalyptus, tea and coffee showed considerable variation both in terms of species diversity and species density. *Eucalyptus* presented a more complex structure as compared to tea and coffee litter, with acarine population always characteristically dominant in eucalyptus litter ranging from 91 to 180/m^2. Tea litter with a very low population of acarines was totally devoid of such insect groups as scarabaeids and thrips.

The collembolan population being higher in the natural forest litter, with their associated fungi, adds to the rapid decomposition. Lull (1959) compared humus depths in virgin and managed system that the depth of humus involved in managed systems was only slightly less than in corresponding virgin stands. The increased occurrence of humus in the virgin stands may possible be due to the increased populations of such microarthropods as acarines and collembolans which increase decomposition.

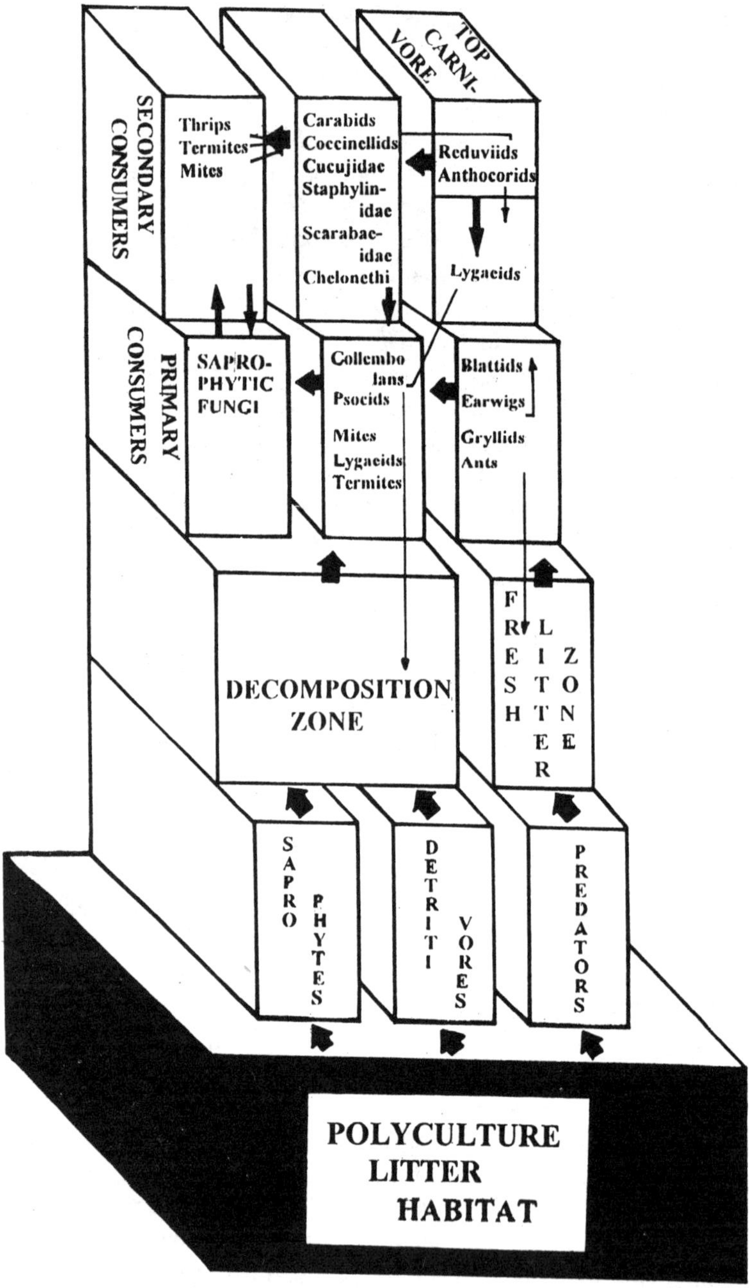

Fig. 35 The polyculture litter habitat.

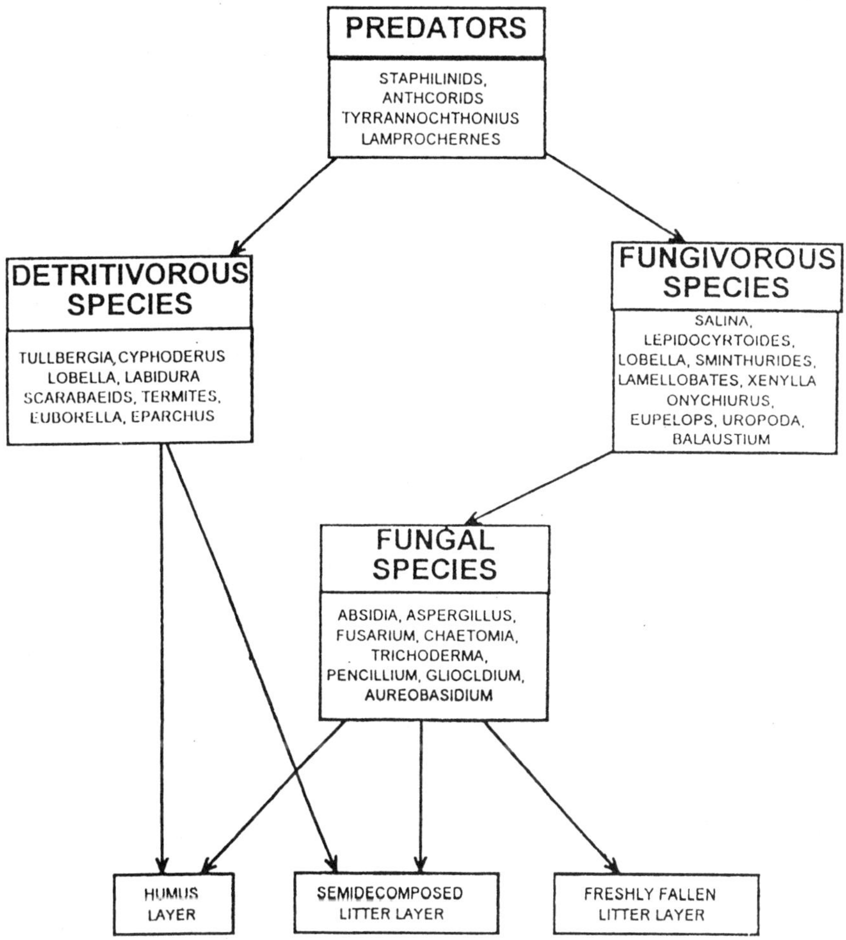

Fig. 36 Food webs and chains associated with monoculture forest-floor litter.

Vertical Stratification of Litter Microarthropods

The heterogenous forest floor comprising as it does an extension of the structural diversity of its vegetation with a complex of biotic and abiotic components, naturally shows alternations in both quality and quantity of the litter. Intact natural forest ecosystems by the diversity of their flora support more diverse decomposer and consumer systems at various trophic levels than those obtained from man-made monoculture practices which result in truncated decomposer and food-chain links resulting in reduced structural complexity. The distribution patterns

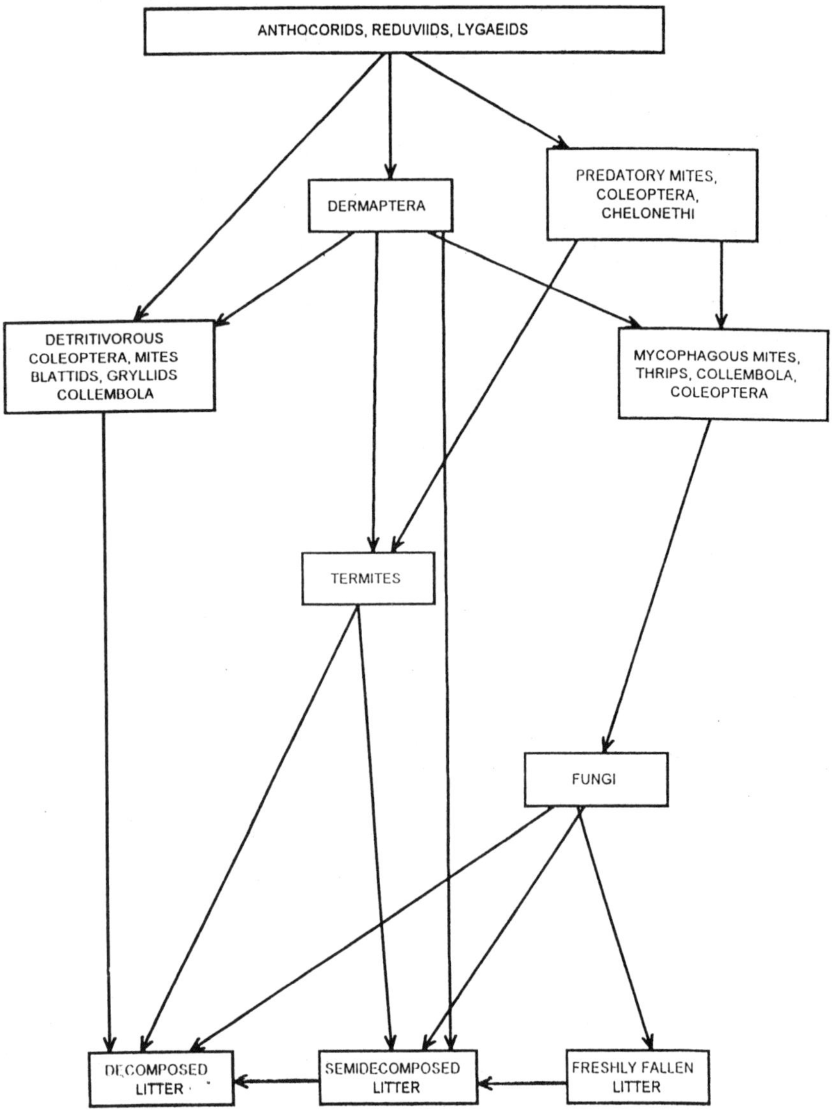

Fig. 37 Interlocking food chains and webs of major fauna and fungi associated with natural polyculture forest-floor litter.

of the microarthropods in terms of vertical distribution of litter also appear important and interspecific competition tends to be reduced by vertical distribution as well as horizontal compartmentalization of resources.

The vertical stratification of microorganisms suitable for a particular microarthropod species or their immatures may be largely responsible for species distribution patterns. Anderson (1971) while explaining the combination of high species diversity with a comparatively low degree of feeding specificity in detritivore communities such as Collembola, suggested that interspecific competition is possible reduced not only by different vertical distribution, but also by the horizontal compartmentalization of resources. The vertical stratification of the collembolan species and the other entomofauna could be due to the vertical stratification of the litter horizons, the nature of the litter, and the microclimatic adaptations of the colonizing species. The abundance of large collembolan forms in the top litter layer and the cryptic forms in the middle and humus zone could also be a behavioural strategy to escape from the predatory arthropods. This differential vertical stratification may be a temporal displacement since many species are not restricted to any one particular layer. Some of the temporal changes in the species composition of populations associated with decomposing litter have been correlated with physical and chemical properties of the litter.

Based on the quality of leaf litter, colonizing insect species vary in the different layers of the litter habitat. Analysis of the vertical components of the litter elements shows the topmost layer comprising the predatory food-chain, with such insect groups as Orthoptera, Hemiptera and Coleoptera being latter represented. The intermediate litter layer, where partially decomposed litter is available, the colonizing insects mainly belong to the orders Thysanoptera, Dermaptera, Coleoptera, and Collembola. The third or transition (humus) layer mainly comprised the collembolans and acarines. In the case of the natural forest litter the collembolans tend to dominate all the other groups, while acarines dominated in monoculture litter.

Individual species composition of collembolans in the three different litter zones (fresh leaf litter, middle and humus zone) as well as at 4 different sampling sites indicated that they form a major component of the entomofaunal complex. Of eight species collected, their occurrence varied in the different litter zones. Species like *Callyntura ceylonica, Salina tricolor* and *S. quatturofasciata* were confined to the layer (ranging from 3 to 65; 4 to 81 and 0 to 43/m^2 respectively). The middle zone is rather complex wih a comparatively high population of such species as *Xenylla* sp., *Hypogastrura* sp., *Cryptophygus thermophylus* and *Tullberiga* sp. dominating all the other collembolans in terms of the population structure. The humus zone is the most important layer which comprised all the species represented in the middle zone, but *Xenylla* sp. appeared to dominate the other collembolans, reaching as high as 2,078/m^2 followed by *Hypogastrura* sp. (410/m^2), *C. thermophylus* (320/m^2), *Tullbergia* sp. (168/m^2) and *Folsomides* sp. (96/m^2). One characteristic feature of this collembolan population structure, is the consistent increase in the population of *Xenylla* sp. and a decline in the population of *Hypogastrura* sp. and vice versa. Temperature differences exist at the different zonations higher temperature ranging from 21°–23°C in the surface litter, with 20°–21°C and 19°–21°C in the middle and humus zones respectively (Fig. 38).

Litter Insect Community Diversity at Elevations

Collembola and mites tend to dominate the insect assemblage in the three litter layers at elevations of 1,440, 1,820 and 2,220 metres in the Shola forests of Kodaikanal Hills, South India. As many as 13 mite species were isolated at 1,440 m of which *Eupterotegaeus ornatissimus, Macrocheles* sp., and *Caloppia basilewskyi* were layer-specific in being present in the fermentation layers. *Galumna* sp. *Uropoda* sp., and *Scheloribates parvus* were the dominant mite species prevalent in all the three layers. *Xylobates capucinus* recorded from F and H layers was site-specific to low elevation. Mite species showed their maximum abundance in the F layer at all the altitudes and layer preference of mites to 1,440 m was clear by their highest species diversity and density at F layer with 13 groups, 9 groups in H layer and only 3 groups in the L layer. Cryptostigmatids with *Galumna* sp., and *Scheloribates* sp., were the major dominant acarine species at all elevations. Species richness at 1,820 m and 2,220 m was lesser with 11 species from the former with site specific *Plateremacus mirabilis* and *Metabelba obtusus* and 9 species groups from the latter. *Macrocheles* sp., *Hypoaspis orientalis, Plateremacus mirabilis, Eupterotegaeus ornatissimus* were restricted to F layer and *Rhysotritia ardua* to the H layer. Mite fauna of the F layer at higher elevations were represented by Trombididae sp., *Oppia* sp., *Phyllhermannia* sp., and *Zetorchestes saltator* as they were absent in L and H layers. Irrespective of the altitudinal variations cryptostigmatids was the dominant order in Shola ranges followed by mesostigmatids with three and prostigmatids with one species. The most dominant species in terms of numbers were the cryptostigmatids *Galumna* sp., and *Scheloribates parvus* and the mesostigmatid *Uropoda discus* as they exhibited high degree of heterogeneity both in terms of abundance and spatial occurrence (Sabu Thomas *et al.* 1994) (Table 19).

Population density of collembolans exhibited marked differences from lower to higher elevation forest litter habitat. All the groups were

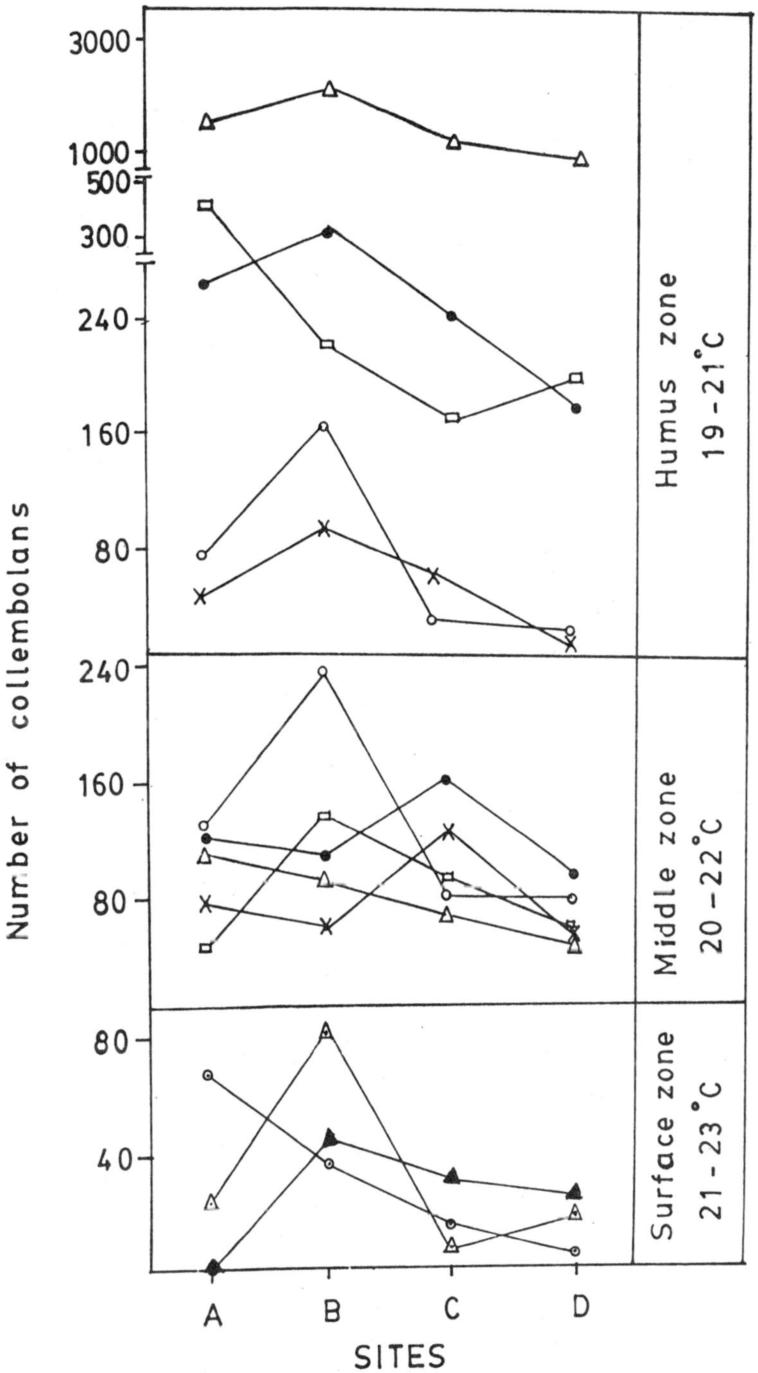

Fig. 38 Distribution of Collembolan species in the natural forest litter at altitude. Δ—*Xenylla* sp., □—*Hypogastrura* sp., ●—*Cryptophygus thermophylus*, ○—*Tullbergia* sp., X—*Folsomides* sp., ⊙—*Callyntrura ceylonica*, ⊙ in triangle—*Salina tricolar*, ▲—*S. quatturofasciata*.

Table 19. Major faunal groups associated with the Shola forests litter at different elevations in the Kodaikanal Hills

Groups	Species/Genera/Family
Acari	*Galumna* sp., *Plateremacus mirabilis, Scheloribates parvus, Uropoda discus, Phyllhermannia* sp., *Xylobates capucinus, Eupterotegaeus ornatissimus, Macrocheles* sp., *Metabelba obtusus, Zetorchestes saltator, Caloppia basilewskyi, Oppia* sp., *Hypoaspis orientalis, Otaheitensis* sp., *Rhysotritia ardua*
Collembola	*Proisotoma* sp., *Isotoma* sp., *Isotomina* sp., *Isotomina thuemophila, Folsomides parvus* sp., *Folsomina* sp., *Folsomides* sp., *Lepidocyrtus* sp., *Pseudosinella* sp., *Seira* sp., *Cyphoderus* sp., *Entomobrya* sp.
Coleoptera	*Carabidae, Scarabaeidae, Staphylinidae, Elateridae, Coccinellidae, Hydroscaphidae*
Thysanoptera	*Stigmothrips, Hoplothrips, Nesothrips, Bactrothrips, Ecacanthothrips, Xylaplothrips, Azaleothrips*
Chelonethi	*Lamprochernes, Geogarypus, Stenatemnus, Tyrrannochthonius, Neobisium*
Formicidae	*Ponera ceylonensis, Leptothorax schurri, Leptothorax* sp.,
Dermaptera	*Euborellia* sp., *Foraminolabis* sp., *Eparchus insignis, Labidura riparia*
Reduviids	*Rhinocoris* sp. *Pirates affinis*
Anthocorids	*Anthocoris* sp., *Xylocoris* sp.

members of Isotomidae and Entomobryidae at each site and exhibited an increase in abundance from L to F layers followed by a fall in the H zone. The F layers at the three elevations showed the maximum species abundance and density with distinct preference for lower elevations. Dominant species at all sites were *Isotoma* sp., *Entomobrya* sp., and *Lepidocyrtus* sp. irrespective of the variation in elevations. Freshly fallen litter along all the elevations was frequented with *Isotoma* sp., *Lepidocyrtus* sp., and *Entomobrya* sp., showing uniformity in their occurrence. *Proisotoma* sp., *Folsomides parvus, Seira* sp., and *Folsomina* sp., and *Isotomina* sp., showed specific preference to the F layer. Of the 11 species recorded at 1,440 m, *Proisotoma* sp., and *Seira* sp., were unique by their absence in higher elevations. Instances of layer specificity is distinct from the F layer specific presence of *Pseudosinella* sp., and *Cyphoderus* sp., at 1,820 m and *Isotoma* sp. *Isotomina* sp., *Folsomides* sp., and *Cyphoderus* sp., at mid-elevation litter matrix. Compared to the abundance of collembolans in the F layer at low and mid-elevations that of higher elevations showed distinct decline. Of the 11 species recorded from Shola litter *Entomobrya* sp., *Lepidocyrtus* sp., *Folsomina* sp., and *Isotomina* sp., have to be considered as highly adaptable and abundant and ecological resilient fauna by their presence in all the three litter layers irrespective of the variations in elevations.

In association with the litter at the three altitudes were the carabids, staphylinids, scarabaeids and Elateridae. While the diversity and richness tended to be similar at low and mid elevations it was highest at 2,220 m indicating altitude specific variations. The dominant coleopteran groups at low and mid elevations was the detritivore complex followed by the fungivore and predatory forms. At 2,220 m predatory carabids and staphylinids were the dominant functional group followed by a reduced presence of detritivore and fungivore coleopterans. It is obvious from a comparison between different elevations, that lower elevations harboured the widest diversity of coleopterans. Though distributional patterns varied significantly between the layers, F layer at all the elevations harboured the maximum number of groups and H layer the maximum of number of grubs whose abundance was higher in the humus layer at 2,220 m. Unlike mites and collembolans, the distribution of coleopterans along the various elevations was uneven and species abundance did not exhibit a linear relationship with species diversity.

Humus layer harboured the maximum number of pseudoscorpions at all the three sites with their maximum abundance at low elevation (30 m^{-2}). Of the five species *Geogarypus* sp., followed by *Amblyolpium* sp., was dominant in H layer and F layer at 1,440 m, whereas *Stenatemnus* sp., gained preponderance at 1,820 m and 2,220 m. In contrast to other species, *Tyrrannochthonius* and *Neobisium* species were absent at higher elevations, whereas at lower elevation their occurrence was in reduced numbers and were strictly confined to the humus layers as indicated by their absence in L and F layers.

Analysis of the litter inhabiting thrips revealed the preponderance of mycophagous species, *viz., Stigmothrips limpidus, Hoplothrips* sp., *Bactrothrips* sp., and *Ecacanthothrips tibialis; Azaleothrips amabilis, Xylaplothrips* sp., and *Nesothrips* sp. Abundance and heterogeneity of thrips was higher at low and mid elevations as represented by the presence of 7 species. *Nesothrips* sp., and *Bactrothrips* sp., were dominant in the L, F layers at all elevations. *Bactrothrips* sp., and *Ecacanthothrips* sp., gained dominance in the surface layer at 2,220 m, though the overall density of thrips community was comparatively meagre at this site. Drastic reduction in the population density was evident in the humus layers as evidenced by their total absence in the H layer at mid and high elevations and the presence of a very reduced number in low elevations.

Species richness and layer preference patterns of other groups, viz., blattids, termites, gryllids and dipterans were similar in all the elevations. Blattids and termites were abundant at lower elevations with a clear preference to the F layer, whereas gryllids were abundant at mid

Table 20. Functional groups of forest-floor arthropodal faunal complexes at two natural forest sites

Functional groups	Order	Representative families/genera at Idikki and Munnar
Predators	Coleoptera	Carabidae, Staphylinidae, Coccinellidae, Cuculidae *Foraminolobis*
	Dermaptera	*Macrocheles, Trombididae, Hypoaspis*
	Acari	*Stenatemnus, Tyrrannochthonius, Amblyolpium,*
	Chelonethi	*Leychitia, Olpium, Calocheiridius, Geogarypus, Lamprochernes*
Herbivores	Coleoptera	Leiodidae, Scarbaeidae, Scolytidae
	Collembola	*Sminthurides*
	Acari	*Scheloribates*
Fungivores	Coleoptera	Ptilidae, Pselaphidae, Leiodidae, Scaphydidae, Nitidulidae, Scolytidae
	Collembola	*Sminthurides, Sinella, Pseudosinella, Hypogastrura, Lepidocyrtus, Isotomurus*
	Thysanoptera	*Bactrothrips, Elaphrothrips, Nesothrips, Xylapothrips, Hoplothrips, Stigmothrips, Ecacanthothrips*
	Acari	*Arcoppia, Scheloribates, Rhynchoribates, Setoxylobates, Megalotocepheus, Apoplophora, Phyllhermannia, Galumna, Eurotocepheus, Zetorchestes*
Detritivores	Coleopterans	Sacrabaeidae, Hydrophilidae, Bostrichidae, Hydroscaphidae, Nitiduliidae
	Dermaptera	*Eparchus, Euborella, Foraminolabis*
	Collembola	*Isotomurus, Hypogastrura, Xenylla*
	Acari	*Arcoppia, Scheloribates, Rhyncoribates, Setoxylobates, Megalotocepheus, Apoplophora, Eurotocepheus, Galumna, Zetorchestes, Phyllhermania*

and high elevations. Exceptions were in the distribution of dipteran flies (44 m^{-2}) and maggots in the L and F layers at low elevations. Of particular interest is their almost total absence from L and F layers at mid and high elevations. However, in contrast to other groups, dermapterans represented by *Foraminolabis* sp., *Eparchus* sp., *Euborellia* sp., and *Labidura* sp., persisted in all the altitudes. *Euborellia* sp., and *Foraminolabis* sp., were predominant at low and mid elevations, whereas at 2,220 m species belonging to *Euborellia* dominated the habitat.

Ant species represented by *Ponera ceylonensis, Leptothorax schurri* and *Leptothorax* sp., were recorded at all the elevations with distinct preference for the F layer as indicated by their abundance, with

Leptothorax schurri as the dominant species. The spec es richness declined with increase in altitude, as the litter zones at 2,220 m recorded the lowest density of ants per square metre of litter.

Top rung of the trophic levels was occupied by the reduviid predators *Rhinocoris* sp., *Pirates affinis* and *Anthocorid* sp., and *Xylocoris* sp., present exclusively in the L layer at 1,440 m and 1,820 m. Of special interest is their absence in the litter matrix at 2,220 m so that in the placement of carabids and rove beetles as the dominant predator fauna followed by Chelonethi and mites (Table 20).

10

Forest Denudation and Guild Structure of Litter Insects

It is a recognised fact that natural forest ecosystems support more diverse consumer and decomposer systems because of their floristic and accompanying faunistic diversity at various tropic levels, than man-made vegetation or forests interfered by man. Reduced structural complexity results from long-term imbalance in interfered forest ecosystems and potentially useful genes are lost when forests are cleared and vegetation cover gets modified leading to a reduction in the stability of the system. The role of forests in stabilizing micro-climate is well known and the soil cover provided by the litter harbours a wealth of insect resources, both according to the degree of litter availability as well as variability, which of course depends on the component tree species. The preservation and protection of forests to safeguard the natural ecosystem as well as the genetic pool they contain, is an important aspect taking into consideration the active role of litter organisms in the breakdown of litter, as well as in sustaining the food-web or trophic complexities in different forest types.

From the point of view of vegetation dynamics, estimates of litterfall are not only important, but also the dynamics of the fauna it harbours. It has earlier been emphasised that leaf contributes more to litter involving around 52–54% of the total litter and may reach even 75% in tropic humid montane forests of South India and estimates indicate that compared to the average total litter over 3 years, leaf litter amounted to 70% (Blasco and Tassy 1975). It has also been suggested that massive defoliation in some years results in the high increase in leaf litter production, alongside with the associated increase in the population of litter arthropod species. Varying quantities of litter are needed for the transformation of a small amount of elements from the plant to the soil, as for instance, 30–35 kg of litter in a tropical rain forest. The highest litterfall recorded is around 23–25 t/ha/yr, with 3.8–7.3 t/ha of leaf litter.

It is well known that tropical forest systems have been subject to considerable human interference. That undisturbed mixed forests have low incidence of pests is well known, with more of the natural regulatory mechanisms involved. In interfered situations, outbreaks are essentially due to climatic stress, absence of predators and parasites or their ineffectiveness.

It may not be out of place to indicate that patterns of deforestation are uneven the world over and as indicated by Myers (1991) several countries are now losing more than 4000 km^2 of moist, tropical forests annually. Besides moist tropical forests, dry tropical forests are also being destroyed. An overall estimate of destruction of both types of forests has been expressed as 40 hectares/minute. The largest single cause of destruction is small-scale farming and accounts for more than 60%, followed by commercial logging and cattle ranching (Myers 1991).

With human interference resulting in the depletion of natural forests and the burning of litter or through typical monoculture practices, both the quantity and quality of leaf litter as well as fungal associates become reduced, naturally followed by a depletion in the tropic chains as well. In plantations treated with insecticides and fungicides there is an increased depletion of leaf litter insects. Variation in the nutrient element in litter is an equally important factor and the abundance of nitrogen varies considerably in litter and nitrogen circulation is often more dominant than other elements, making, therefore, the need for a study of the seasonal variation of ash content of leaf litter very useful.

In tropical forests most of the nutrient stock is in the biomass and nutrients in litter falling in the dry season tend to be rapidly mineralised and taken up by roots in the wet season (Dickinson and Pugh 1974). In this connection it may be indicated that cellulose, lignin, hemicellulose, water soluble sugars, amino acids and aliphatic compounds are essentially involved in plant litter. While it is known that earthworms digest a good proportion of cellulose and chitin because their gut enzymes contain cellulase and chitinase, there is collective degradation by many organisms, so that decomposition of litter and soil faunal activity goes hand in hand. Blattids, earwigs, carabids, collembolans, pauropods, symphylids and diplopods and a host of insects play a role in the mechanical fragmentation of litter. Subsequently microbes, bacteria and fungi play a significant role in litter breakdown.

Human interference with forest ecosystems in general and tropical rain forests in particular has therefore caused widespread concern regarding the consequences of interference which have been known to affect climate, biodiversity and human societies. While biomass is created by photosynthesis in virgin stands there is a reversal of the process during forest denudation, with the loss of litter naturally resulting in poor soil formation, reduction or loss of biodiversity, particularly the loss of ecotypes and other genetic variation. Extensive denudation will particularly affect biological control resources some of which occur at low densities, which would be vulnerable to genetic erosion or chance extinction with their reduced numbers. This will naturally affect the components of the ecosystem and the very few survivors can hardly support or sustain the trophic web.

In an earlier chapter stress was made of the role of mineralization as an important process for the replenishment of nutrient pool. The combined action of the decomposer community comprising the microorganism and macro- and microarthropods results in a change of organic matter in molecules. Phenols, tannins, lignins and enzymes are known to play a vital role in decomposition. The volatile chemicals emanating from the litter through various stages of breakdown has been known to activate many insects, and with the loss of litter such integrated interaction among the litter insect community becomes very much reduced or even lost. Human activities like burning, increasing removal of litter including branches for firewood have had their serious effects.

It has to be reemphasized that from the surface litter layer to humus and the soil horizon below, the soil environment is one of the most dynamic sites of biological interactions in nature, in which several biochemical reactions occur concerned in the decomposition of organic matter or in mineral recycling. The organic portion of the soil, viz., the humus therefore results from the synthetic and decomposing activities of the microflora and fauna. Humus exists in a dynamic state and being in a continuous state of activity, leads to the loss of carbonaceous material and at the same time new microbial tissue is generated. The rate of carbon loss can be related to the structure and fertility of the soil and also reflects the level of biological activities. Humus contains a number of polymerized substances, aromatic molecules, polysaccharides, basic amino acids and phosphorus-containing compounds and as indicated earlier, from the surface litter or 'L' layer through the F and H layers, a host of volatile chemicals are released varying with the composition of litter. It is this chemical diversity of litter resources that is responsible for sustaining the biological diversity of litter with the varying communities of micro- and macroarthropods becoming involved in spatial and temporal relationships. The impact of human interference therefore has considerable bearing on litter insect guilds resulting in their fragmentations and ultimate disappearance, positively affecting species diversity, not to mention loss of soil fertility.

Ecological studies relating to forested and deforested sites in West Bengal, have shown marked variation in total arthropod fauna and species diversity of Collembola as compared with the resource forest floor. Principal factors involved relates to the comparatively higher temperature in the deforested sites and in the reserve forested sites of the forest floor, the increased moisture content stimulate decomposition, ammonification, and nitrification through the bacterial activity resulting in higher population. Variation in the faunal make up is possibly due to the adverse ecological conditions in the deforested sites, which showed significantly low content of organic carbon and nitrate resulting in the poor soil environment and hence reduced fauna. Interestingly enough, the collembolan fauna represented by several species, four of them, namely *Salina indica, Isotemniella minor, Isotomina interrupta* and *Entomobrya* sp. occur in a wide variety of habitats showing high habitat tolerance or ubiquitous nature. Euryoecious and stenoecious species of Collembola are typical of both reserved and deforested sites (Table 21).

Periodic burning of forests at many places in the Western Ghats involving the removal of accumulated litter not only results in the substantial loss of source of organic matter in the forest floor, but also alters the vegetative composition of the undergrowth and affect the faunal community. Comparison of fire-prone and fire-free sites shows a definite reduction in the density of faunal elements.

Table 21. Euryoecious and stenoecious Collembola at natural and denuded forest sites

Collembolan species at natural forests	Collembolan species at denuded forest sites	(Ubiquists) Collembola present at both sites
Megalothorax minimus	*Isotomina thermophila*	*Salina indica*
Lepidocyrtus sp.	*Lepidocyrtus suborientalis*	*Isotemniella minor*
Folsomoides exigua	*Cyphoderus* sp.	*Isotomina interrupta*
Xenylla obscura	*Proisotoma* sp.	*Entomobrya* sp.
Pseudochorules sp.	*Isotomodes* sp.	
Hypogastrura sp.	*Willemia* sp.	
Cyphoderus javanus	*Arrhopalites* sp.	
Dicyrtomina sp.	*Sminthurides* sp.	
Alloscopus tetracantha		
Sinella sp.		
Lepidocyrtus scaber		
Siera sp.		
Lobella sira		
Lobella maxillaris		
Sminuthrides appendiculatus		
Tullbergia sp.		
Subisotòma fitchioidis		

(after Hazra 1991)

The curtailment of litter influx after clearance is the main control on nutrient inputs to and the resultant soil fertility of cleared sites. The significance of leaf litter as a nutrient reserve and as protection against soil erosion and the importance of the need to retain leaf litter soil organic matter in the soil after deforestation cannot be overemphasized and if soil erosion is to be avoided. Ecological engineering has been suggested in the redevelopment of the disturbed forest zooamelioration, and to result in important functioning of soil organisms, especially the macrofauna in order to promote turnover in the whole production cycle. Zooamelioration therefore aims at improvement of life condition and stimulation of the local soil animal population. Hence, reconstruction of forest bioresources is a new approach.

In conclusion, it may be indicated that there is some beneficial effort of soil invertebrates particularly in ecosystems with a higher seasonal patterns of litterfall and this benefit accrues from the fact that their biomass also acts as a reservoir of plant nutrients (Ananthakrishnan 1986). The saprovores consume litter and are, in turn, consumed by predators. Increased populations of litter invertebrates, ensure a wide range of mortality, so that the nutrients contained in their tissues become mineralised slowly. This is confirmed by the fact that the tissues of the litter feeding invertebrates have higher concentration of sodium, phosphorus and nitrogen. Physical and chemical properties of litter contribute towards temporal changes in species composition of population associated with decomposing litter. Certain microarthropods could be used as indicator species for particular stages of decomposition such as a bacterial phase, succeeded by an Acari phase and finally a humification phase involving collembolans, thrips, oribatids, pauropods, symphylids etc.

Litter/soil fauna influence texture and soil consistency, water movement and retention of soil chemical and nutritional properties. Culture practices that affect soil fauna adversely are mechanical methods of land clearing, use of fire as a management tool, indiscriminate use of agrochemicals and monoculture. Tropical soils devoid of macro-micro elements are vulnerable to compaction and structural collapse and rapid soil degradation. Deposition, removal and physical destruction affect litter types and leaching and biological activity may transform one type of litter into another facilitating or interfering with decomposer and consumer activity. The relative importance of each of these processes in determining the balance of litter at a given site varies with the ecosystem (Facelli and Pickett 1991). In short, it is the effects of litter on the abiotic environment that directly or indirectly influences the effects of litter on living organisms.

It has also to be emphasised that food-webs and interactions between organisms also require intensive study, as very little is known of them in the major subecosystems like canopy, subcanopy, forest floor, soil strata etc., so as to assess their role. With Zooamelioration techniques, population of soil/litter organisms tend to recover and the rate at which resistance occurs presumably depends on the length of the food-chain, which again depends on web feature such as number of predator and prey species etc., long food-chain tend to reduce the rate at which population densities recover from environmental disasters (Pimm 1991).

11

Basic Techniques in the Study of Chemical Ecology of Litter

Insect faunal elements inhabiting the litter on the soil surface in very large numbers have to be carefully isolated both for purposes of determination and assessment of populations. To get an idea of homogeneity of distribution it is necessary to take several samples from nearby localities which are apparently similar. There is a need to take samples of undecomposed litter, F and H layers together. At least ten sampling sites each of a square metre demarcated by frames of a square metre size and sampled material must be treated as soon as possible to prevent escape of the faunal elements. From the chemical ecological viewpoint, each sampling site will have a different composition of litter elements particularly in a natural forest, which will provide diversity in the chemical composition, notably of volatile chemicals.

It would not be out of place to mention that extractions from litter samples are carried out by the dry Tullgren funnel method, with slight modifications (Murphy 1962). Funnel sizes may vary, from 15 to 22.5 cm diameter at the top, and generally of polythene or zinc, with a lid covering the top through which an electric wire passes to suspend an electric bulb of approximately 15–25 watts. The surface of the litter sample gets warmed and care should be taken to see that it is not warmed to more than 40°C. About 3–8 days are usually needed to dry the sample and to enable extraction of large number of samples simultaneously, a series of Tullgren funnels are mounted on a wooden frame (Fig. 39) with a series of bulbs for heating. The length of the exit tube of the funnels varies according to the size of the funnel and it dips into a small container with 75% alcohol in which the microarthropods drop. The bulb is placed at least 10 cm above the level of the sieve on which rests the litter mass.

While larger or macrofaunal elements are usually handpicked, through a combination of floatation and funnel extractions, Acarina and Collembola can be effectively recovered. When the litter is heated slowly

Fig. 39 Tullgren funnels.

so that the sample temperature is around 30°C, with a maximum up to 38°C, all the Collembola and Acarina tend to leave the litter, besides thrips, pseudoscorpions, psocids etc. However, temperature effect varies with species and a range of temperatures appears to have a definite influence on the microarthropods.

Litter Insects and Chemical Environment

Primary and secondary metabolites are equally typical of the litter environment, with an array of phenolics and volatiles, all of which contribute towards a better understanding of litter insect communities. This is very evident from the diversity of chemical substances which vary for monoculture to polyculture litter and even among the polyculture litter the type of plant litter and age of the litter have important bearings on the chemical ecology. Besides the capacity of insects to perceive chemical compounds, there is increasing evidence to show that the interacting effects of several chemical stimuli are more important than individual stimuli and enable aggregation of individuals in a litter environment and the volatiles released essentially determine their behavioural diversity.

The overall concentration of amino acids, proteins, carbohydrates, lipids and nitrogen, tends to vary with the age and type of litter. Available nitrogen and the concentration of minerals such as calcium, magnesium, sodium and potassium, besides water play a very important role in the distribution of litter organisms. In view of the role of fungi in the decomposition of litter, relevant techniques used in the culture of fungi are included. Since chemical ecology of litter plays a major role in the distribution and activities of several insect faunal elements, a basic idea of the techniques involved has been presented.

I. Biochemical Estimations of Plant Litter

1. Moisture Content

- — Weigh 2 g of fresh plant litter material
- — Keep in incubator at 37°C till a constant weight is obtained
- — The difference between the initial and final weight is calculated and moisture content represented in percentage

2. Total Proteins (Lowry *et al.* 1951)

Regents

5% TCA	—	Prepared by dissolving 5 g of TCA in 100 ml of distilled water
0.1 N NaOH	—	4 g in 1,000 ml distilled water
Solution A	—	2 g sodium carbonate, 400 mg sodium hydroxide in 100 ml distilled water
Solution B	—	500 mg copper sulphate, 1 g potassium tartrate, in 100 ml distilled water
Solution C	—	50 ml solution A + 1 ml solution B

- — Weigh 500 mg of sample, grind using a pestle and mortar and homogenize with 20 ml of 5% TCA
- — Centrifuge and disslove residue in 5 ml of 0.1 N NaOH
- — Pipette 0.1 ml of the above and make up to 1 ml with distilled water
- — Add 5 ml of solution C
- — After 10 minutes add 0.5 ml of Folin Ciocalteau (1:1 reagent)
- — Read OD at 750 nm after 10 minutes
- — Bovin Serum Albumin is used as standard

3. Total Carbohydrates (Dubois *et al.* 1956)

Reagents

5% TCA

80% Phenol
Concentrated sulphuric acid
Glucose (AR)

- Weigh 500 mg sample, grind using a pestle and mortar and homogenize in 20 ml of 5% TCA
- To 0.1 ml of supernatant add 0.05 ml of 80% phenol
- Add 5 ml of concentrated sulphuric acid with the stream of acid being directed against the sample surface
- Allow tubes to stand for 10 minutes, shake and after 20 minutes read OD at 490 nm
- Glucose is taken as standard

4. Total Nitrogen (Vogel 1963)

Reagents

Potassium sulphate
Copper sulphate
Concentrated sulphuric acid
50% sodium hydroxide
4% Boric acid
0.1N Hydrochloric acid

- Take 100 mg of sample in a clean dry kjeldal flask
- Add 5 g of potassium sulphate, 0.25 g of copper sulphate
- Pour 25 ml of concentrated sulphuric acid in such a way as to wash down any solid adhering to the neck
- The flask is heated slowly until solution becomes colourless
- Transfer completely to the ammonia distillation apparatus
- Add few fragments of porcelain, 50% NaOH. The ammonia evolved is distilled off to a solution of 4% Boric acid
- The borate formed is determined by titration with a standard 0.1 N HCl (1 mole of HCl being required for each mole of ammonia)

5. Total Lipids

Reagents

Petroleum ether

- Weigh 500 mg of sample and grind with petroleum ether
- Filter to a previously weighed glass tube
- Homogenise residue again with petroleum ether and filter
- After 4–5 such repeated extractions, the filtrate is placed in an oven at 50°C for evaporation
- Reweigh the tubes; the difference in weight gives the amount of lipids present

6. Total Amino Acids (Moore and Stein 1948)

Reagents

4N Sodium citrate buffer of pH 5.0 (20.5 ml of 0.2 molar citric acid plus 29.5 ml of 0.2 molar sodium citrate)
Ninhydrin reagent
— Grind 100 mg of sample in 10 ml of buffer and centrifuge
— 0.1 ml of sample is made to 1 ml with buffer and 1 ml of ninhydrin is added
— Place tubes in boiling water bath for 15 minutes
— Cool, dilute with 3 ml of water
— Read OD at 570 nm
— DL-alanine is used as a standard

7. Protein Profiles

Polyacrylamide Gel-Electrophoresis

For qualitative profiles of proteins polyacrylamide gel electrophoresis (PAGE) can be used based on the methodology adopted by Davis (1964). The detailed procedure is as follows:

The following 3 stock solutions should be prepared, filtered and refrigerated in dark coloured bottles for long-term regular use.

Solution A

1 N HCl	=	48.0 ml
Tris (Hydroxymethyl amino methane)	=	36.6 g
N,N,N′,N′, Tetramethyl ethylene diamine (TEMED)	=	0.046 ml
Glass distilled water	=	100 ml
pH	=	8.8

Solution B

Acrylamide	=	30 g
Crystallised N. Methylene bisacrylamide (Bis)	=	0.8 g
Glass distilled water	=	100 ml

Solution C

Ammonium persulphate	=	140 mg
Glass distilled water	=	100 ml

Buffer solution for reservoirs

Tris salt	=	6 g
Glycine	=	28.8 g
Glass distilled water	=	1,000 ml
pH	=	8.6

As much as 100 ml of the stock solution should be made up to 1 litre with glass distilled water and used as tank buffer. The bottom of upper container will have 12 holes (9 mm diameter) at equal intervals along the circumference, 2 cm from the periphery. Each hole encloses a rubber grommet of 5 mm (ID) which permits a water-tight seal when the gel tubes are subsequently inserted. After polymerization of gels in the glass tubes, samples (200 µg of each) in 10% sucrose, were layered directly on top of the gel. After removing the rubber caps carefully, it is filled with Tris-glycine buffer. After filling the upper and lower containers with 500 ml of Tris-glycine buffer, a few drops of tracer dye (Bromophenol blue 0.01% aqueous) should be added in the upper tank. The current can be adjusted to give 2–2.5 MA per tube for about 2.5 hr until the tracer dye migrates to a distance of about 90–95 mm into the running gel. Electrophoresis should be performed in a temperature range of 5°–10°C.

Protein Staining and Destaining of Gels

The gel columns should be removed from the respective tubes by rimming them with a blunt needle in cold distilled water. Immediately after their removal they can be stained with 0.2% Comassive brilliant blue G 250 in a mixture of methanol, acetic acid, water (25 : 7 : 68 v/v). After staining for 24 hr, the gels should be destained with the same solvent mixture without the stain until the bands are very clear.

Scanning the Gels

The gels can be scanned in a LKB Bromma Laser Densitometer (Model 2202).

Preparation of Working Solution

Electrophoresis can be carried out with the small pore gel tubes having a concentration of 7.5%.

7.5% Polyacrylamide Gel Electrophoresis

Solution A	=	3 ml
Solution B	=	6 ml
Solution C	=	12 ml
Water	=	3 ml

Acrylamide Polymerization

As many as 12 open-ended glass tubes (4 mm ID × 100 mm long) should be stopped with rubber caps at one end placed with the open end up. The running gel solution should be pipetted into each tube to a depth of 90.95 mm, taking care to avoid air bubbles with a pasteur pipette. Glass distilled water should be slowly added to avoid mixing with gel solution. A sharp refractive bounding will be visible between the gel solution and water. The entire set-up should be kept undisturbed at 28°C for 30 minutes. After polymerization, the water layer can be removed with the help of a pasteur pipette.

Electrophoresis

Electrophoresis can be carried out in an apparatus having two perspex circular tanks (14/10 cm). Each tank will contain platinum electrode placed vertically in the centre.

The Rm values of the protein fractions as calculated by the formula:

$$\text{Rm} = \frac{\text{Distance travelled by the protein fraction}}{\text{Distance travelled by the tracer dye}}$$

from the Rm value, various protein fractions can be identified with the help of electropherogram.

II. Use of Thin Layer Chromatography (TLC) and UV Spectrophotometry in the Characterisation and Quantification of Allelochemics in Litter Samples

Thin layer chromatography enables the identification and quicker separation of plant litter chemicals with accuracy and high sensitivity and resolution. UV spectrophotometry is an added advantage to TLC for characterisation and quantification of plant chemicals. The UV spectrum of a compound is generally measured over the 200–400 nm range, although the visible range up to 900 nm may also be scanned by the instrument. Quartz cells are transparent over this range and compounds are dissolved in a solvent such as 95% ethanol for analysis. UV spectrophotometry is a non-destructive technique, in that samples may be recovered if necessary for further analysis.

Thin Layer Chromatography Principle (TLC)

Thin layer chromatography involves the separation of compounds on glass plates coated with a thin layer of supporting material. This stationary phase may operate on either the partition, adsorption, ion-exchange mode. The majority of the separations in TLC are carried out on silica gel, where the effect is mainly due to partition, as in the case of paper chromatography. The method is thus rapid and has a high resolving power.

a. Preparation of TLC plates

The glass plates to be coated are cleaned thoroughly, free of any grease and wiped with alcohol and dried in special racks. They are lined on the applicator pad with a few drops of water used as a means for dicing these plates immobile on the pad.

Coating materials

The weighed absorbent (50 g/100 ml) is mixed with water either by grinding in a mortar or by vigorous shaking for 10 minutes in a stoppered conical flask. Later it is transferred quickly to the spreader to have an uniform layer.

Coating the plates

The aperture-opening in the applicator is set for desired thickness of the coating (usually 25µ). The slurry is poured into the applicator placed on the last plate on the left of the pad. The lever is turned 180°C. When the slurry could be seen coming out, the applicator is drawn smoothly on the plates.

Drying and storing

The coated plates are adequate for many separations and are left in position overnight free from dust. When necessary, these plates are activated after drying in a hot air oven at 105°C for 30–60 minutes. The plates are placed in special racks provided, to prevent the layers from chipping off. Since these active plates are easily deactivated in moisture, they are stored in a desiccator till use.

b. Application of the Samples

A small spot is applied to the silica gel plates and chromatograms are developed in various solvent systems.

UV Spectrophotometry: A spectrophotometer is a sophisticated type of colorimeter where monochromatic light is provided by a grating or prism. The band width of the light passed by a filter is quite broad, so that it may be difficult to distinguish between two compounds of closely related absorption.

The white light emitted from the light source passes through the filter F and slit S and goes to the monochromator utilizing a concave diffraction grating. In the monochromator, the light is converted into a monochromatic light. It passes through the exit slit S2 and mirror M2, and reaches the rotating mirror M3 where it is divided into two light beams; one is used for reference, and the other is for sample. These two light beams, after passing through the sample compartment, alternatively irradiate the R 928 ultra-high sensitivity detector where they are converted into two electric signals.

The model Hitachi 150-20 ultraviolet spectrophotometer is used for measurement of absorption or transmittance of liquid, solid and gaseous samples in ultraviolet, visible and near-infrared regions. This function is effective in reading photometric data for quantitative analyses at an optionally selected wavelength. The display indicates the photometric function for wavelength spectra and scans wavelengths to display and record sample spectra. The display is a measuring function which indicates how photometric values changes with the lapse of time. It is used for displaying and recording the change in photometric values and automatically printing out the values changing with time with a colorimeter.

c. Separation of Phenols and Flavonoids from Hydrolysed Plant Extracts by Thin Layer Chromatography

A general procedure for surveying plant litter for phenols and flavonoids has been evolved in the laboratory based on a modification of a method developed by Bate-Smith (1962).

Extraction procedure for individual flavonoids

(1) A small amount of plant litter is immersed in 2N HCl and heated in a test tube for 30–40 min. at 100˚C. (2) The cooled extract is then filtered and extracted with ethyl acetate. (3) When the solution is found to be coloured (either because the original litter is coloured with anthocyanin or because colour has formed from leucoanthocyanidin during acid treatment) the aqueous extract is further heated to remove the last trace of ethyl acetate and re-extracted with a small volume of amyl alcohol. (4) The ethyl acetate extract is concentrated to dryness, taken up in 1–2 drops of ethanol and aliquots chromatographed one dimensionally, alongside authentic markers, in five solvents: Forestal (Acetic acid–con. HCl–Water, 10 : 3 : 30), 50% aqueous acetic acid, Butanol–Acetic acid–Water; 4 : 1 : 5. (5) The amyl alcohol extract is concentrated to dryness and then taken up in a few drops of 1% methanolic HCl and aliquots chromatographed in Forestal and Formic acid–con. HCl–Water (5 : 2 : 3). (6) Then, the compounds which had eluted from TLC plates are easily detected, since they give characteristic fluorescent colours in UV light, which are intensified by further treatment with ammonia vapour. Identification is confirmed by spectral measurements using UV spectrophotometer.

Separation Procedure for Individual Phenols

(1) Plant litter (fresh or dried) is extracted in aqueous alcohol. (2) Acid hydrolysis with 2N HCl for 0.5 hr. (3) Cooled and filtered. (4) Alkaline hydrolysis with 2N NaOH at room temperature. (5) Phenols transferred to ether and the ether extract should be evaporated and dried. (6) Residue redissolved in ether and chromatographed one dimensionally, alongside authentic markers, in three solvents; acetic acid and chloroform (1 : 9), ethyl acetate and benzene (9 : 11) and benzene, methanol and acetic acid (45 : 8 : 4).

III. Application of HPLC in the Preliminary Identification of Nutritional Compounds and Phytochemicals Active against Litter Insect Community

The High Pressure Liquid Chromatography (HPLC) is an excellent analytical tool for the assay of phytochemicals of plant origin. This technique is very useful for the isolation of free fatty acids, phenols and amino acids in as much as it is non-destructive to the natural chemical compounds during the course of analysis. The HPLC systems closely follow the principles of open-column systems and also facilitates the assay of large amount of biologically active phytochemicals by a biologist. Microcolumns are available with 3 μm packing with less solvent use and with faster separation. New techniques in precolumn derivatization has enabled us recently to lower the detection limit in a compound mixture hence requiring low quantity of samples. The ability of HPLC to monitor difference in concentrations of bioactive chemicals between samples will greatly simplify studies and enable us to monitor and quantify bioactive compounds in various developmental stages of plants or insects.

1. Chromatography Principle

Chromatography is a technique by which structurally similar substances are separated by a continued redistribution between two phases, one stationary and the other mobile. In HPLC, the mobile phase is pumped through the column by high pressure. Therefore, HPLC is sometimes called as High Pressure Liquid Chromatography. Chromatographic separations of biomolecules are developed based on differences in the physico-chemical nature of the compounds. These include size, charge, affinity and hydro-phobicity. A variety of attractive forces between the stationary phase and the substances to be separated have been recognised and their mode of action provides some insight into the

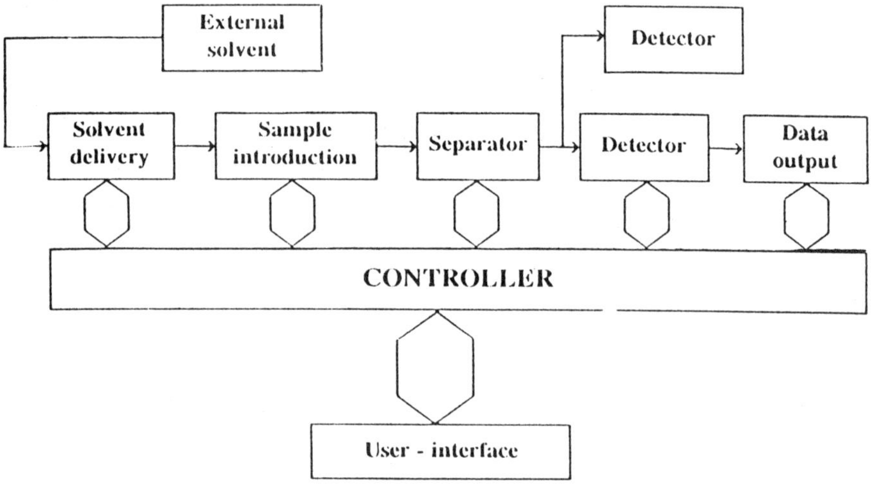

Fig. 40 Schematic diagram indicating the principal components in a HPLC.

mechanisms responsible for separation in HPLC. The degree of polarity (solvents, solutes and absorbents) can be roughly related to the number of bands which a group can form. An asorbent can be polar, such as silica gel and alumina, or they can be non-polar, such as charcoal or Bondapak C_{18}. Both adsorption and partition separations depend upon the polarity differences between the solutes being separated (Fig. 40).

2. General Functional Aspects of HPLC

The system consists of a reservoir from which solvent passes at a constant rate through a filter to an injection unit. The component in the sample become distributed differently between the mobile and stationary phases because they have differential interactions (physical and chemical) with each phase. The components move with different speeds through the column, depending on their affinity for the different phases and so they elute from the column at different times (known as retention time). By changing the composition of mobile phase during the elution, it is possible to analyse a wide variety of compounds in a given time that would be possible under constant conditions.

After the separator, the column eluent enters the detector, which measures the physical and chemical property of each eluent. The profile of peaks which result from an injected sample comprise the chromatogram. Peaks are usually identified by their retention time,

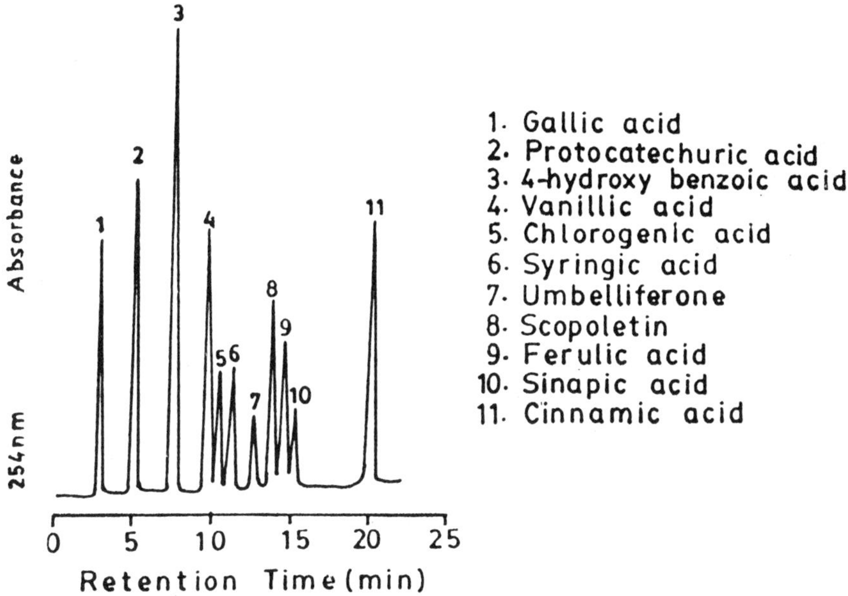

Fig. 41 HPLC chromatogram of phenolics.

which is the time required to elute the corresponding band from the column (Fig. 41).

3. Sample Preparation for HPLC Analysis

Sample preparation for HPLC separation of bioactive compounds depends on the type of sample to be separated. The following factors are taken into consideration before sample preparation.

a) Solution compatible with the model of HPLC selected
b) Free from interfering matrix elements
c) Stable compounds
d) Inappropriate concentration for detection and measurement

Fatty acid methyl esters

For analysing the fatty acid content in litter samples, the extracted lipids of litter should be taken and added with 10 ml of ethanol, 3 ml of 28% ammonium hydroxide, 25 ml of petroleum ether and 25 ml of diethly ether to a separating funnel, shaken for 5 minutes and allowed to stand for 20 minutes. The bottom phase should be drained off, and the upper ether phase is dried, to which 3 ml of 0.5N NaOH in methanol is added and heated in a steam bath for 15 minutes. To this 5 ml of water and 2N HCl should be added slowly until the pH reaches approximately 2. The fatty acid methyl esters can then be extracted into 5 ml of petroleum ether and 5 ml of diethyl ether from acidified methylated lipid extract.

4. Selection of Columns and Packings for HPLC

A column usually consists of a cylindrical stainless steel tube (2–8 mm ID) which holds the stationary phase particles in place. Columns are available in different lengths and diameters. The type of column to be used depends on the physical and chemical properties of a sample. Silica gel is generally used as a base because it is not crushed under high pressure in HPLC, due to its pore size and distribution.

a. Fatty acids can be analysed by Hewlett Packard High Performance Liquid Chromatograph system (Model HP 1090) at 210 nm, using Hypersil ODS 5 μm column with water and acetonitrile as mobile phase at a flow rate of 0.45 ml according to the gradient programme of Schuster (1985). The retention times and area per cent of fatty acid methyl esters can be recorded.

b. For analytical separation of anthocyanins and flavonoids by Hewlett Packard HPLC system a slight modification in columns and analytical condition is required. A good separation of these polyphenolic compounds can be achieved in a Nucleosil 5 μm column (C 18) with (A) formic acid/water (5/95 v/v) and (B) methanol as solvents at a flow rate of 0.5 ml/min following the gradient elution technique.

5. Detectors Used in HPLC

Detectors utilise one or more physical characteristics of the sample being separated. The following detectors are used in HPLC separation:

* UV detector
* Fluorescence detector
* Refractive index detector
* Electrochemical detector

The HPLC system currently available at the Institute is provided with an UV detector which holds good for detection of variety of compounds.

6. Gradient Elution Technique in HPLC

The HPLC columns may be run isocratically, i.e. with a constant composition eluent or they may be run in the gradient elution mode in which the mobile phase composition varies during the run.

IV. Identification of Potential Volatile Compounds in Diverse Litter Samples

The gas chromatography offers one of the most versatile techniques for analysing complex mixtures of volatile substances from plant litter and animal sources. The use of plant litter volatiles in host location and host selection process by litter insect community is an important aspect. In this respect GC/MS becomes a very valid tool to determine the volatile chemical distribution within and between the litter layers for effective exploitation in the study of nature of insect colonization among the plant litters.

1. Gas Chromatography Principle

The gas chromatography is a special form of the general chromatographic technique, in which the mobile phase is a gas and the stationary phase may be either solid or liquid. The technique is suitable for separation of materials which are volatile without decomposition.

In gas chromatography, the sample is introduced into the moving gas stream and is carried by it through the column. The active solid or non-volatile liquid forms the stationary phase, whereas the carrier gas forms the mobile phase. The components of the sample mixture distribute themselves between the two phases. The absorption or solubility in the fixed phase tends to differ from substance to substance, therefore these components are carried along the column at different rates and finally emerge at the end of the column in distinct zones (peaks) separated by the carrier gas. On emerging, the vapours of compounds are detected by a suitable detector with an automatic recording device (Fig. 42).

2. General Functional Aspects of Gas Chromatography

A carrier gas like nitrogen, hydrogen, helium, argon or a mixture of any two of them, which is obtained from a cylinder is passed through a pressure adjusting valve, drying tube and flow regulator to give an even flow of gas and then to the column. The gas then passes through the sample injection point where it picks up the sample for analysis.

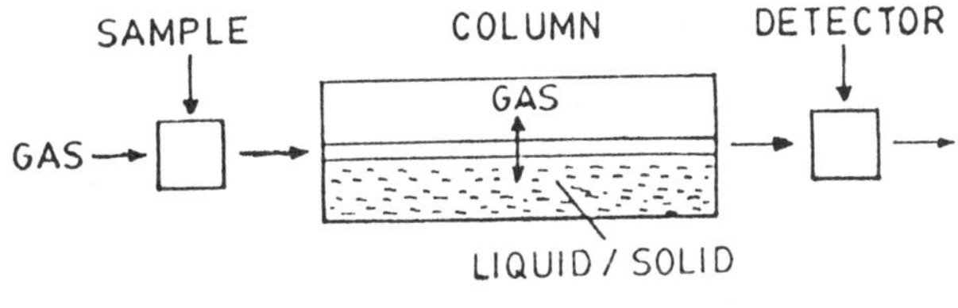

Fig. 42. Basic principle of gas chromatography.

The carrier gas containing the sample enter into the column where they are separated. The gas chromatography (GC) columns are made of glass, metal or a plastic tubing in coiled or uncoiled form and is placed in a temperature controlled chamber (known as oven) where a constant temperature or a programmed temperature is maintained. The carrier gas containing the separated vapours of the compounds emerge from the column and pass through the detector. The detector measures the concentration of the vapour in the carrier gas or its total amount. Different type of detectors are used and each detector has a selected advantage for accurate identification and quantification of separated compounds. The general working procedure of the detectors provide an electrical signal which is proportional to the concentration of the vapours in the carrier gas. The signals from the detector reaches an electronic recorder which provides the chromatogram of the concerned compound. If necessary, arrangements could be made to collect the gas stream through a sample collector for identification and for further specific analysis (Fig. 43).

3. Sample Preparation and Volatile Analysis in GC-MS

The litter sample selected for volatile terpenes are washed repeatedly in hexane and the solvent evaporated under vacuum. The concentrated hexane fraction is injected as described in the methodology section into a Hewlett Packard GC with a mass spectral detector. Fused silica capillary column (10 m × 0.2 m) with a crosslinked methyl silicone phase is used. Helium is used as the carrier gas which is regulated by a flow meter. The GC/MS system is interfaced with a chemstation with a chemical library (NBS49K) containing mass spectra of over 49,000 compounds. The temperature regime is maintained at 40° to

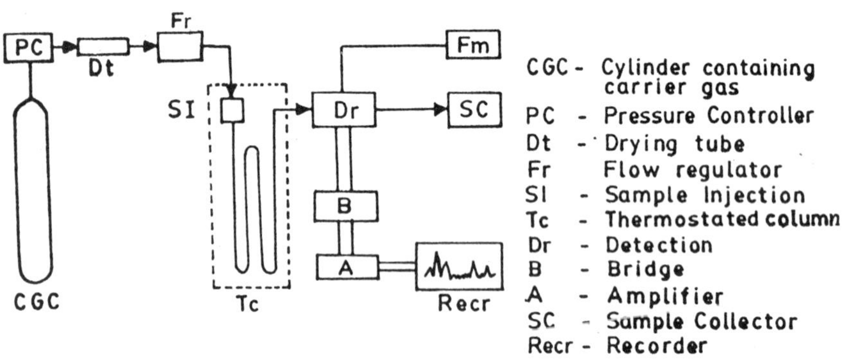

Fig. 43 Flow chart indicating various sections in a gas chromatograph.

250°C at 5°C rise per minute with a 2 minute solvent delay. Injector transfer line and ion sources are set at 220°C and 230°C respectively. The chromatograms obtained during the assay is analysed using a chemstation NBS49K library for final compound identity.

4. Some important Applications in Gas Chromatography

The type of columns used for compound separation play an important role in the efficiency of the instrument. Above all, some functional conditions of the instrument is very vital for the effective separation. Let us examine some of the crucial factors important in volatile separation using a gas chromatograph. The columns are not normally operated at room temperature except for the analysis of gaseous mixtures or extremely volatile liquids and therefore it must be maintained at a suitable elevated temperature. The oven in the column chamber takes care of constant temperature or maintain the programmed temperature for a good volatile separation. Use of vapour jackets in column chamber maintains column temperature to the accuracy of 0.1°C. This aspect is considered as a very important step because alteration in column temperature may affect the normal volatile separation.

The sensitivity of any gas-chromatographic assembly depends upon the sensitivity of detectors used in the study. The detection system consists of three main parts: (a) the detector itself usually gives the electric output; (b) the associated instrument for display of the signal from the detector; and (c) in most of the cases, the automatic recorder which gives a finished chromatogram. The major function of the detector is the signal when the output gas is carrying and not carrying the foreign molecules. Once this is known it is easy to calibrate the output reading by feeding samples of known composition into the instrument.

5. Study of a Chromatogram

A chromatogram is a plot of the detector response which measures the changes of composition of the column effluent against time of the volume of the carrier gas. The chromatogram may be of two types, viz. differential and integral, depending on whether it measures the instantaneous concentration in the effluent gas in the total amount of sample accumulated from the beginning of the analysis. If a chromatogram is obtained automatically, the horizontal axis is the direction of the motion of the recorder on the paper. It also represents the volume of the carrier gas which has passed if the flow rate is constant. For differential chromatogram the vertical axis represents the concentration of the solute in the gas.

6. Study of a Mass Spectrum

A mass spectrum is a presentation of the masses of the positively charged fragments versus their relative concentration, the most intense peak in the spectrum, called the base peak, is assigned a value of 100%, and the intensities of other peaks are reported as percentages of the base peak. A graphic representation has the advantage of presenting patterns that with experience can be quickly recognised.

7. Computer Coupling with Gas Chromatographs

Modern gas chromatographs as it is available at this Institute (Hewlett Packard, 5890 series II) is coupled with a computer which is of great help in several respects. They are: (1) The computer attachment saves valuable time in terms of giving operative commands to the GC as well as to the detector. (2) The mass spectral data from the detector enters the computer for final compound identification. The computer is equipped with a mass spectral library which includes several thousand volatile compounds and act as a reference. (3) The computer package also provides facilities to store data on several separations for a holistic study. Finally, computer data analysis system quicken the entire working system of the gas chromatograph and its detectors.

V. Flame Photometry

Principle

If a solution containing a metallic salt (or some other metallic compound) is aspirated into a flame, a vapour which contains atoms of the metal may be formed. Some of these gaseous metal atoms may be raised to an energy level which is sufficiently high to permit the emission of radiation characteristic of that metal as shown in Fig. 44.

Preparation of Standard Solutions

1. Sodium, Potassium and Magnesium

Weigh accurately 2.542 g AR quality of sodium chloride (NaCl) and transfer it into a one litre volumetric flask through a funnel.

Weigh accurately 1.909 g AR quality of potassium chloride (KCl) and 0.809 mg of $MgCl_2$ and transfer them into the same volumetric flask through the same funnel. Add double distilled water to the flask, dissolve the crystals and make up the solution to the mark. This stock

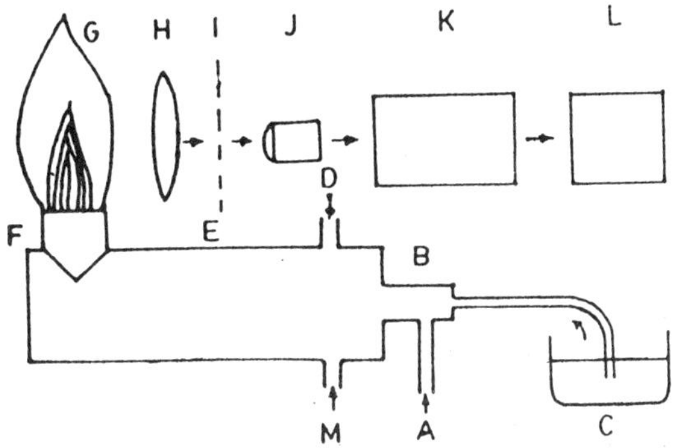

A : AIR
B : ATOMIZER
C : SOLUTION OR SAMPLE
D : GAS
E : MIXING CHAMBER
F : BURNER
G : FLAME
H : LENS SYSTEM
I : FILTER
J : DETECTOR
K : AMPLIFIER & SIGNAL PROCESSOR
L : READOUT
M : DRAIN

Fig. 44 Flame photometer—principle of operation.

standard solution contains 1,000 ppm/1000 ppm of sodium (Na), potassium (K) and magnesium (Mg).

This stock standard solution is successively diluted further with double distilled water to have working standard solutions of lower concentrations of 100, 50, 20, 10, 5, 2 and 1 ppm, of Na, K and Mg.

2. Calcium

Weigh accurately 0.2497 g of AR quality calcium carbonate ($CaCO_3$) and dissolve it in dilute hydrochloric acid (HCl), transfer it into a one litre volumetric flask and make up the solution to the mark with double distilled water. This stock standard solution contains 100 ppm of calcium (Ca).

The stock standard solution is successively diluted further with double distilled water to have working standard solutions of lower concentrations of 50, 20, 10, and 5 ppm of Ca.

Sample preparation

As much as 5 g of incinerated litter sample was dissolved in 100 ml of diluted hydrochloric acid. This solution is then transferred to one litre volumetric flask and made up to 1 litre by adding double distilled water. This sample is taken and used for the mineral analysis.

Analysis

The sample thus prepared is taken and introduced into the flame photometer. The sample is then sprayed over the gas flame and the excitation of the molecules is carried out under carefully controlled and reproducible condition. The derived spectral line is isolated using appropriate filters and the intensity of the light (OD) which is approximately equal to the concentration of the element is measured in the detector. The OD value of the samples were compared with these standards and the concentration of them in volumetrics into sample were calculated.

VI. Litter Microbial Community

The physical and chemical environments influence litter accumulation and litter biomass may be reduced by physical and chemical degradation, heterotrophic consumption and decomposition by microfaunal and microbial elements. Further, the degree of litter breakdown as evident from the increased activities of associated microbial elements is also an important facet of litter ecosystem. The litter microbial species can be isolated by the following techniques.

Analysis of Litter Microbial Communities

For analysis, the litter microbial communities can be cultured in the laboratory using any one of the following methods.

(a) Moisture Chamber Culture Method

The microbial element is cultured by placing the litter particles on moistered filter papers in petri dishes. The colonies emerging from the media were later separated out to make semi-permanent slides with individual fungi for identification.

(b) Litter Dilution Technique

One gram of litter sample is taken and shaken well in 100 ml of double distilled water. One ml of this aliquot from each litter suspension is inoculated into petri dishes containing 20 ml of Czapek-Dox agar medium at 10^{-3} dilution. The inoculum is then incubated at room temperature for 2 to 3 days and the number of colonies that emerge is considered as representing the population size of fungi in litter.

(c) Potato-Dextrose-Agar Plating Method

(i) Preparation of sample

For analysis of microflora, one gram of representative litter sample is to be washed with sterile distilled water and plated onto a potato-dextrose-agar medium for culturing the spores.

(ii) Plating

Sterile, moulten, potato-dextrose-agar medium was poured into sterile petri plates with a few drops of 30% lactic acid under aseptic condition. Acidifying the medium with lactic acid was to avoid bacterial contamination. The sample containing the microflora of the litter was plated onto the medium after it had set and left at 34 ± 2°C for 2–3 days for the emergence of colonies.

Calculation

Total populations of fungal colonies can be calculated as mean number of fungal colonies × dilution factor/weight of litter sample.

References

Abe, T. 1982. Ecological Role of Termites in a Tropical Rain Forest. In: *The Biology of Social Insects.* Eds. Bread, M.D., Michener, E.D. and Evans, H.E. Westview Press, Boulder, Colorado, 71–75 pp.

Aber, J.D. and Mellilo, J.M. 1982. Nitrogen immobilization in decaying hardwood leaf litter as a function of initial nitrogen and lignin content. *Can. J. Bot.* 60: 2263–69.

Ananthakrishnan, T.N. 1973. *Mycophagous tubulifera of India.* Occl. Publ. Ent. Res. Int. Loyola Coll. Madras, 144 pp.

Ananthakrishnan, T.N. 1979. Microarthropods and soil ecosystems. *J. Bombay Nat. Hist. Soc.* 75(3): 625–631.

Ananthakrishnan, T.N. 1981. Thrips. In: *Vectors of Plant Pathogens.* Eds. K.F. Harris and K. Maramorosch. Academic Press, New York, 149–164 pp.

Ananthakrishnan, T.N. 1984. *Bioecology of Thrips.* Indira Publishing House, Michigan, 233 pp.

Ananthakrishnan, T.N. 1986. *Bioresources Ecology.* Oxford and IBH Publishing Co. Pvt. Ltd., New Delhi, 172 pp.

Ananthakrishnan, T.N. 1988. Dynamics of the litter ecosystem in natural and interfered forests. *Int. J. Ecol. Environ. Sci.* 14: 61–66.

Ananthakrishnan, T.N. 1990. *Reproductive Biology of Thrips.* Indira Publishing House, Michigan, 158 pp.

Ananthakrishnan, T.N. and Suresh, G. 1983. Patterns of fungal resource utilisation and feeding range in some mycophagous Tubulifera. *Proc. Indian Acad. Sci.* (Anim. Sci.). 92(4): 285–293.

Ananthakrishnan, T.N. and Sabu K. Thomas 1993. Chemical ecology of two heterogenous forest litter ecosystem and related trophic interation of insect communities. *Int. J. Ecol. Environ. Sci.* 19: 143–160.

Ananthakrishnan, T.N., Gopichandran, R. and Subramanian, V.R. 1993a. Chemical ecology of litter breakdown in *Tectona grandis* and its impact on the dynamics of natural insect communities at varying altitudes. *Int. J. Ecol. Environ. Sci.* 18: 63–74.

Ananthakrishnan, T.N., Gopichandran, R. and Subramanian, V.R. 1993b. Dynamics of microarthropod communities in relation to the chemical environment of decomposing *Eucalyptus globulus* litter. *Int. J. Ecol. Environ. Sci.* 19: 83–92.

Ananthakrishnan, T.N., Padmanabhan, B., Dhileepan, K. and Suresh, G. 1984. Ecological interactions, species dynamics and reproductive biology of some Mycophagous Thysanoptera. *Occl. Public. Ent. Res. Inst.* Loyola College, Madras, 5: 1–37.

Anderson, J.M. 1975. Succession, diversity and trophic relationship of some soil animals in decomposing litter. *J. Anim. Ecol.* 44: 475–95.

Anderson, J.M. 1978. Competition between two unrelated species of soil Cryptostigmata (Acari) in experimental microcosms. *J. Anim. Ecol.*. 47: 787–803.

Anderson, J.M., 1988. Invertebrate-mediated transport processes in soils. *Agric. Eco Envir.* 24: 5–19.

Anderson, J.M. and Bignell, D.E. 1980. Bacteria in the food, gut content and faeces of the litter feeding millipede *Glomeris marginata* (Villers). *Soil Biol. Biochem.* 12: 251–254.

Anderson, J.M. and Swift, M.J. 1983. Decomposition in Tropical Forests. In: *Tropical Rain Forest, Ecology and Management.* Eds. Sutton, S.L., Whitmore, T.C., and Chadwick, A.C. Oxford Blackwell Sci. 287–309 pp.

Anderson, R.V., Coleman, D.C. and Lole, C.V. 1983. Effects of Saprotrophic Grazing on Net Mineralization. In: *Terrestrial Nitrogen Cycles.* Eds. Clark, F.E. and Rosswall, T. *Ecol. Bull.* (Stockholm) 33: 201–216.

Anderson, J.M., Ineson, P.K. and Huish S.A. 1983a. The effects of animal feeding activities on element release from deciduous forest litter and soil organic matter. pp. 87–100. In: Proc. VIII Int. Colloq. Soil Zool. Louvain-Laveure. Eds. Lebrum, P.H., Andre, H.M., De Medts, A., Gregoire-Vilso, C. and Waughty, G.

Anderson, J.M., Ineson, P.K. and Huish, S.A. 1983b. Nitrogen and cation mobilisation by soil fauna feeding on leaf litter and soil organic matter from deciduous woodland. *Soil Biol. Biochem.* 15: 463–467.

Anderson, J.M., Proctor, J. and Vallack, H.M. 1983. Ecological studies in four contrasting lowland rain forests in Gunny Malu national Park, Sarawak III. Decomposition and nutrient losses from leaf litter. *J. Ecol.* 71: 503–527.

Annadurai, R.S., Chandrasekar, S.S. and Balu, A. 1988. Trophic structure and diversity in some litter inhabiting microarthropods of monoculture and natural forest ecosystems. *Proc. Indian Acad. Sci.* (Anim. Sci.). 97(4): 301–308.

Aoki, J.I. 1967. Microhabitat of Oribatid mites in forest floor. *Bull. Nat. Sci. Mus. Tokyo.* 10: 133–138.

Appel, H.M. 1993. Phenolics in ecological intractions: The importance of oxidation. *J. Chem. Ecol.* 19: 1543–1551.

Attiwill, P.M. 1968. Loss of elements from decomposing litter. *Ecology* 49: 142–145.

Bate-Smith, E.C. 1962. Attractants and Repellants in Higher Animals. In: *Phytochemical Ecology*. Ed. Harborne, J.B. Academic Press, London, 45–56, pp.

Behan, V.M. and Hill, S.B. 1978. Feeding habits and space dispersal of Oribatid mites in the North American Arctic. *Rev. Ecol. Biol. Sol.* 15: 497–516.

Bengtsson, G., Erlandsson, A. and Rundgren, S. 1988. Fungal odour attracts soil collembola. *Soil Biol. Biochem.* 20: 25–30.

Bengtsson, G., Hedlund, K. and Rundrgen, S. 1991. Selective odor perception in coil Collembola *Orychiurus armatus*. *J. Chem. Ecol.* 17(11): 2113–2125.

Birch, L.C. and Clark, D.P. 1953. Forest soil as an ecological community with special reference to the fauna. *Quart. Rev. Biol* 28: 13–35.

Blasco, F. and Tanny, B. 1975. E'tdude d'un e'cosysteme forester montagnard du sud de l' India. *Bull. E'col.* 6: 525–539.

Bocock, K.L. 1963. Changes in the amounts of dry matter, nitrogen, carbon and energy in decomposing woodland leaf litter in relation to activities of soil fauna. *J. Ecol.* 51: 555–566.

Bocock, K.L. 1964. changes in the amounts of dry matter, nitrogen, carbon and energy in decomposing woodland leaf litter in relation to the activities of the soil fauna. *J. Ecol.* 52: 273–284.

Bocock, K.L. and Gilbert, O. 1957. The disappearance of leaf litter under different woodland conditions. *Pl. Soil* 9: 179–185.

Borkott, H. and Insam, H. 1990. Symbiosis with bacteria enhances the use of chitin by the springtail *Folsomia candida* (Collembola). *Biol. Fertil. Soils* 9: 126–129.

Bradow, J.M. 1991. Relationships between chemical structure and inhibitory activity of C6 through C9 volatiles emitted by plant residues. *J. Chem. Ecol.* 17(11): 2193–2212.

Bradow, J.M. and Connick, W.J. Jr. 1990. volatile of seed germination inhibitors from plant residue. *J. Chem. Ecol.* 16: 645–666.

Bray, J.R. and Gorham, E. 1964. Litter Production in Forests of the World. In: *Advances in Ecological Research.* Ed. Cragg, J. Academic Press, 2: 101–157.

Brown, S. and Lugo, A.E. 1982. The storage and production of organic matter in tropical forests and their role in global carbon cycle. *Biotropica.* 14: 161–187.

Brown, S. and Lugo, A.E. Biomass of tropical forests: A new estimate based on forest volumes. *Science* 233: 1290–1293.

Brunig, E.F. 1973. Biomass Diversity and Biomass Sampling in Tropical Rain Forest. In: *IUFRO Biomass Studies.* Ed. Young, H.E. Orno. Univ. of Mame, 269–293 pp.

Butcher, J.W., Srider, R. and Srider, R.J. 1971. Bioecology of edaphic Collembola and Acarina. *Ann. Rev. Entomol.* 6: 249–288. Stanford. Palo Alto.

Buxton, R.D. 1981. Changes in the composition and activities for termite communities in relation to changing rainfall. *Oecologia* 51: 371–378.

Champion, H.G. and Seth, S.K. 1968. A revised survey of forest types of India. FRI. 122 Government of India, New Delhi, 404 pp.

Choudhuri, D.K. and Banerjee, S. 1975. Qualitative and quantitative composition of Acari and Collembola in relation to soil organic matter-microbes complex. *Orient. Ins.* 9: 313–316.

Choudhuri, D.K. and Pande, T. 1979. High altitude soil animals and their relation with soil factors, with special reference to mites. *Rev. Ecol. Biol. Sol.* 16(2): 219–226.

Christiansen, K. 1964. Bionomics of Collembola. *Ann. Rev. Entomol.* 9: 147–178.

Clement, J.L., Lloyd, H., Nagnan, P. and Blum, M.S. 1989. n-tetradecyl propionate: Identification as a sex pheromone of the eastern termite *Reticulitermes flavipes. Sociobiology* 15: 19–24.

Cole, D.W. and Rapp, M. 1981. Elemental Cycling in Forest Ecosystems. In: *Dynamic Properties of Forest Ecosystems.* Ed. Reichle, D.E. 341–409 pp. CUP, M.

Cole, C.V. Elliot, E.T., Hunt, H.W. and Coleman, D.C. 1978. Trophic interactions in soils as they affect energy and nutrient dynamics. V. Phosphorus transformations. *Microb. Ecol.* 4: 381–387.

Coleman, D.C. and Mc Ginni, J.T. 1970. quantification of fungus-small arthropods food chains in the soil. *Oikos* 21: 134–137.

Coleman, D.C., Reid, C.P.P. and Cole, C.V. 1983. Biological Strategies of Nutrient Cycling in Soil Systems. In: *Advances in Ecological Research 13.* Eds. Macfadyen, A. and Ford, E. Academic Press, New York, 1–55 pp.

Coulson, C.B., Davies, R.I. and Lewis, D.A. 1960. Polyphenols in plant, humus and soil. 1. Polyphenols of leaves, litter and superficial humus from mull and more sites. *J. Soil Sci.* 11: 20–29.

Crossley, D.A. 1977. the Roles of Terrestrial Saprophagous Arthropods in Forest Soils: Current Status of Concepts. In: *The Role of Arthropods in Forests Ecosystems.* Ed. Mattson, W.J. Springer-Verlag, New York, 49–56 pp.

Crossley, D.A. Fr., and Hoglund, M.P. 1962. A litter-bag method for the study of microarthropods inhabiting leaf litter. *Ecology* 43: 571–573.

Cunnigham, R.K. 1963. The effect of clearing a tropical forest soil. *J. Soil Sci.* 14(2): 334–345.

Dadd, R.H. 1981. Essential Fatty Acids for Mosquitoes, other Insects and Vertebrates. In: *Current Topics in Insect Endocrinology and Nutrition.* Eds. Bhaskaran, G., Friedmann, and Rodriquez, I.G., Plenum Press, New York, 189–214 pp.

Davis, B.J. 1964. Disc electrophoresis II. Method and application to human serum proteins. *Ann. New York Acad. Sci.* 121: 404–427.

Dickinson, D.H. and Pugh, G.J.F. 1974. Biology of plant litter decomposition. Vols 1 and 2, Academic Press, New York.

Dindal, D.L. 1977. *Biology of Oribatid Mites of Syracus.* State University, New York.

Dinsdale, D. 1974. Feeding activity of phthiracarid mite. (Arachnida: Acari). *J. Zool.* 174: 15–21.

Doran, J.W. 1980. Soil microbial and biochemical changes associated with reduced tillage. *Soil Sci. Soc. Am. J.* 44: 765–771.

Dubois, M., Cilles, K.A., Hamilton, J.K., Rebers, P.A. and Smith, F. 1956. Calorimetric determination of sugars and related substances. *Anal. Chem.* 28: 351–356.

Durham, K.M. 1989. Litterfall, nutrientfall and production in an *Acacia albida* woodland in Zimbabwe. *J. Trop. Ecol.* 5: 227–238.

Edwards, C.A., Reichle, D.E. and Crossely, D.A. 1970. The Role of Soil Invertebrates in Turnover of Organic Matter and Nutrients. In: *Analysis of an Ecosystem. The Temperate Deciduous Forest.* Ed. Reichle, D.E. Berlin: Heidelberg, New York. Springer.

Einhellig, F.A. 1987. Interactions Among Allelochemicals and other Stress Factors of the Plant Environment. In: *Allelochemicals Role in Agriculture and Forestry.* Ed. Walker, G.R. ACS symposium series 330. American Chemical Society. Washington DC, 343–357 pp.

Everaerts, C., Bonnard, O., Pasteds, J.M., Roisin, Y. and Konig, W.A. 1990. (+) – x – Pinene in the defensive secretion of Nasutitermes princeps. *Experientia* 227–230 pp.

Faber, J.H. and Verhoef, H.A. 1991. Functional difference between closely related soil arthropods with respect to decomposition process in the presence or absence of pine tree roots. *Biol. Biochem.* 23: 15–23.

Facelli, J.M. and Picket, S.A.A. 1991. Plant litter: Its dynamics and effects in plant communities structure. *Bot. Rev.* 57: 1–32.

Feller, M.C. 1980. Biomass and nutrient distribution in two eucalyptus forest ecosystem. *Aust. J. Ecol.* 5: 309–333.

Forest, W.G. and Ovington, J.D. 1971. Organic matter changes in age series of *Pinus radiata* plantations. *J. Appl. Ecol.* 7: 177–186.

Foster, R.C. and Martin, J.K. 1981. In situ Analysis of Soil Components of Ecological Origin. In: *Soil Biochemicals.* Vol. 5. Eds. Paul, E.A. and Ladd, J.N. Mared Dekker me NY. Dasel, 75–111 pp.

Fox, L.R. and Macauley, B.J. 1977. Insect grazing on Eucalyptus in response to variation in leaf tanins and nitrogen. *Oecologia* 29: 145–162.

French, R.C. 1985. The bioregulatory action of flavour compounds on fungal spores and other propagules. *Ann. Rev. Phytopathol.* 23: 173–179.

Gadgil, R.L. and Gadgil, P.D. 1975. Suppression of litter decomposition by mycorrhizal roots of *Pinus radiata. N.Z.J. For. Sci.* 5: 33–41.

Gadgil, R.L. and Gadgil, P.D. 1978. Influence of clearfelling or decomposition of *Pinus radiata. N.S.J. For. Sci.* 8: 213–224.

Gilbert, O.J.W. and Bocock, K.L. 1960. Changes in leaf litter when faced on the surface of soils with contrasting humus types. II. Changes in the nitrogen content of Oak and ash leaf litter. *J. Soil Sci.* 11: 10–19.

Gill, R.W. 1969. Soil microarthropod abundance following old field manipulation. *Ecology* 50: 805–816.

Golley, F.B. 1983. The Abundance of Energy and Chemical Elements. In: *Tropical Rainforest Ecosystems.* Ed. Golley, F.B. Elsevier, Amsterdam, 101–105 pp.

Gossley, (Jr.), D.A. 1977. The Role of Terrestrial Saprophagous Arthropods in Forest Soils. Current Status of Concepts. In: *The Role of Arthropods in Forest Ecosystems.* Ed. Mattson, J. Springer-Verlag, NY, pp. 47–56.

Griffiths, D.K. and Bowman, C.E. (Eds) 1984. *Acarology* VI. Vol. I. Ellis Horwood Limited, John Wiley and Sons, 645 pp.

Grubs, P.J. 1977. control of forest growth and distribution on wet tropical mountains w.s.r. to mineral nutrition. *Ann. Rev. Ecol. Syst.* 8: 83–107.

Gupta, S.R. and Rajvanshi, R. 1991. The Effect of Biotic and Abiotic Factors on Litter Decomposition in a Tropical *Dalbergia sissoo* Forest Ecosystem. In: *Advances in Management and Conservation of Soil Fauna.* Eds. Veeresh, G.K., Rajagopal, D. and Viraktamath, C.A., USA Tech Series, 491–496 pp.

Guru, B.C. and Panda, S. 1991. The Role of *Cryptostyges thermophilus* in Regulating C/N Ratio. In: *Management and Conservation of Soil Fauna*. Eds. Veeresh, G.K., Rajagopal, D. and Viraktamath, C. Oxford and IBH Publ. Co. Pvt. Ltd., 99–107 pp.

Hagvar, S. 1991. Succession of Microarthropods in Decomposing Spruce Needles: A Litter Bag Study over Ten Years. In: *Advances in Management and Conservation of Soil Fauna*. Eds. Veeresh, G.K., Rajagopal, D. and Viraktamath, C. Oxford and IBH. NA., 485–490 pp.

Hale, H.W. 1965. Observations on the breeding biology of Collembola-I and II. *Pedobiologia* 5: 146–152; 161–177.

Hale, H.W. 1967. Collembola. In: *Soil Biology*. Ed: Burges, A. and Raw, F. Academic Press, 397–411 pp.

Hanlon, R.D.G. and Anderson, J.M. 1979. The effects of Collembola grazing on microbial activity in decomposing litter. *Oecologia* 38: 93–99.

Hansley, W.R.C. 1954. Mull and mor formulation in relation to forest soils. *Bull. For. Comm. Lond.* 23.

Haq, M.A. and Konikkara, I.D. 1988. Microbial Associations in Xylophagous Oribatids. In: *Progress in Acarology,* vol. 1. Eds. Channabasavanna, G.P. and Viraktamath, C.A. Oxford and IBH Publ. Co. Pvt. Ltd., New Delhi, 469–474 pp.

Hazra, A.K. 1978a. Effects of organic matter and water content of soil on the distribution of Collembola (insecta) in an uncultivated field of West Bengal. *Bull. Zool. Surv. India.* 1(2): 107–114.

Hazra, A.K. 1978b. Ecology of Collembola in a deciduous forest floor of Burdwan district (W.B.) in relation to soil moisture. *Oriental Insects.* 12: 265–274.

Hazra, A.K. 1991. Effect of Deforestation on the Soil Macro and Microarthropod Fauna of West Bengal, India. In: *Advances in Management and Conservation of Soil Fauna*. Eds. Veeresh, G.K., Rajagopal, D. and Viraktamath, C.A. Oxford and IBH Publishing Co. Pvt. Ltd., New Delhi, 399–412 pp.

Hillis, W.E. 1966. Variations in polyphenol composition within species of *Eucalyptus*. L'herit. *Phytochemistry.* 5: 541–556.

Hirose, T. 1975. Relations between turnover rate, resource utility and structure of some plant populations: A study of the matter budgets. *J. Fac. Sci. Univ. Tokyo* III. XI(II): 355–407.

Hobbie, S.E. 1992. Effects of plant species on nutrient cycling. *Trec* 7(10): 336–339.

Holdridge, L.R., Grenke, W.C., Hatheway, W.H., Liang, T. and Tosi, J.A. 1971. *Forest Environments in Tropical Life Zones—A Pilot Study*. Pergamon, Oxford.

Haffaker, C.B. 1958. Experimental studies on predation: disposal factor and predator–prey oscillation. *Hilgardia* 27: 343–383.

Ineson, P., Leonard, M.A.. and Anderson, J.M. 1982. Effect of collembolan grazing upon nitrogen and cation leaching from decomposing leaf litter. *Soil Biol. Biochemistry* 14: 601–605.

Jain, N.K. 1963. The effect of decomposition of mature leaves and litter of Sal (*Shorea robusta*) on garden and lateritic oil. *Proc. Nat. Acad. Sci.* 33: 148–152.

Janzen, D.H. 1986. The future of tropical ecology. *Ann. Rev. Ecol Syst.* 17: 305–324.

Jenny, H., Gessel, S.P., Bingham, F.T. 1949. Comparative study of decomposition rates of organic matter in temperate and tropical regions. *Soil Sci.* 68: 419–432.

Jordan, C.F. and Kline, J.R. 1972. Mineral cycling: Some basic concepts and their application in a tropical rain forest. *Ann. Rev. Ecol. Syst.* 3: 33–50.

Katase, T. 1981. Distribution of different forms of p-hydroxybenzoic, Vanillic, p-Coumaric and ferulic acids in forest soils. *Soil Sci. Plant Nutr.* 27: 365–371.

Klinge, H. 1978. Litter production in tropical ecosystems. *Malay. nat.* 30: 415–422.

Krucheva, G.F. 1960. the role of invertebrates in the decomposition of oak litter. *Soviet Soil Sci.* 4: 360–365.

Kshattriya, S., Sharma, G.D. and Mishra, R.R. 1992. Enzymes activities related to litter decomposition in forests of different age and altitude in North East India. *Soil Biol. Biochem.* 24: 265–270.

Kuhnelt, W. 1963. Soil inhabiting Arthropoda. *Ann. Rev. Entomol.* 8: 115–136.

Kuhnelt, W. 1976. *Soil Biology.* Faber and Faber, London, 483 pp.

Kurup, V.C. and Prabhoo, V.R. 1981. Some Aspects of Reproduction in *Cyphoderus javanus.* In: *Progress in Soil Biology and Ecology in India.* Ed. Veeresh, G.K. UAS. Tech Series No: 37. UAS, Bangalore, 120–124 pp.

Lahdesmaki, P. and Pushpanen, R. 198. Degradation products and the hydrolytie enzyme activities in the humification processes. *Soil Biol. Biochem.* 20: 287–292.

Lathwell, D.J. and Grove, T.L. 1986. Soil-plant relationships in the tropics. *Ann. Rev. Ecol. Syst.* 17: 1–16.

Leavings, S.C. and Windsor, D.M. 1985. Litter arthropods population in a Tropical deciduous forest: relationship between years and arthropod groups. *J. Anim. Ecol.* 54: 61–69.

Lee, K.E. 1991. The Role of Soil Fauna in Nutrient Cycling. In: *Advances in Management and Conservation of Soil Fauna.* Eds. Veeresh,

G.K., Rajagopal, D. and Viraktamath, C. Oxford and IBH Publ. Co. Pvt. Ltd., New Delhi, 465–472 pp.

Lee, K.E. and Wood, T.G. 1961. *Termites and Soils.* Academic Press, New York, London, 251 pp.

Leigh, E.G. Jr. 1975. Structure and climate in tropical rain forest. *Ann. Rev. Ecol. Syst.* 6: 67–86.

Leonard, M.A. 1984. Observations on the influence of culture conditions on the fungal feeding preferences of *Folsomia candida. Pedobiologia* 26: 361–367.

Lewis, N.B. and Harding, J.H. 1963. Soil factors in relation to pine growth in South Australia. *Aust. For.* 27: 27–34.

Ling-lee, M., Ashford, A.E. and Chilvers, G.A. 1977. A histochemical study of polysaccharide distribution in eucalyptus mycorrhizas. *New Phylol.* 78: 329–335.

Longman, K.A. and Jenik, J. 1987. *Tropical Forest and its Environment.* John Wiley and Sons, New York.

Longstaff, B.C. 1980. the functional response of a predatory mite and the nature of the attack rate. *Aust. J. Ecol.* 5: 151–158.

Lowry, O.H., Rosenbrough, N.G., Farr, A.L. and Randall, R.G. 1951. Protein measurement with Folin phenol reagent. *J. Biol. Chem.* 193: 265–275.

Lussenhop, J. 1992. Mechanisms of Microarthropod-Microbial Interactions in Soil. *Adv. Ecol. Res.* 23: 1–33.

Luxton, M. 1972. Studies on the oribatid mites of a Danish beach wood soil. I. *Nutr. Biol. Pedobiol.* 12: 434–63.

MacCauley, B.J. 1975. Biodegradation of litter in *Eucalyptus panciflora* communities. I. Techniques for comparing the effects of fungi and insects. *Soil Biol. Biochem.* 7: 341–344.

Madge, D.S. 1965. Leaf-fall and litter disappearances in tropical forest. *Pedobiologia* 5: 273–288.

Maldagive, M.E. 1964. Importance des populations de termites dans les sols eqvatorianx. Trans. 8th. Int. Congr. Soil Sci. Bucharest. 3: 743–745.

Manton, S.M. 1977. *The Arthropoda.* Clarendon, Oxford.

Mary, M.V. and Sankarn, K.V. 1991. Ex-situ decomposition of leaf litters of *Tectona grandis, Eucalyptus tereticornis* and *Albizzia falcataria.* KFRI Research Report 71. 41 pp.

Mc Bryer, J.F., Reichle, D.E. 1971. Trophic structure and feeding rates of forest soil invertebrate populations. *Oikos* 22: 381–88.

Meentemeyer, V. 1978. Macroclimate and lignin control of litter decomposition rates. *Ecology* 59: 465–472.

Meentemeyer, V., Box, E.O. and Thompson, R. 1982. World patterns and amounts of terrestrial plant litter production. *Bioscience* 32: 125–128.

Mellilo, J.M., Aber, J.D. and Murabore, J.M. 1982. Nitrogen and lignin control of hardwood leaf litter decomposition dynamics. *Ecology* 63: 621–626.

Misra, R.A. 1972. A Comparative Study of Net Primary Productivity of Only Deciduous Forests Open Land of Varanasi, India. In: *Trcpical Ecology Emphasising Organic Production*. Eds. Follen, E.B. and Golley, P.M. University of Georgia, Atlens, 279–283 pp.

Mitchell, M.J. 1978. Vertical and horizontal distributions of oribatid mites (Acarina: Cryptostigmata) in an aspen woodland soil. *Ecology* 59: 561–525.

Mitchell, M.J. and Parkinson, D. 1976. Fungal feeding of oribatid mites (Acarina: Cryptostigmata) in an aspen woodland soil. *Ecology* 57: 302–12.

Molina, A., Reigosa, M.J. and Carballeira, A. 1991. Release of allelochemical agents from litter, throughfall and topsoil in plantations of *Eucalyptus globulus* Labill in Spain. *J. Chem. Ecol.* 17(1): 147–160.

Moore, S. and Stein, W.H. 1948. Photometric ninhydrin method for use in the chromatography of aminoacids. *J. Biol. Chem.* 176: 367–388.

Mound, L.A. 1974. Spore-feeding thrips (Phlaeothripidae) from leaf litter and dead wood in Australia. *Aust. J. Zool.* 27: 1–106.

Mound, L.A. 1976. American leaf-litter Thysanoptera of the genera Erkosothrips, Eurythrips and Terthrothrips (Phlaeothripidae: Phlaeothripinae). *Bull. Br. Mus.* (*Nat. Hist.*) *Ent. Jr.* 35: 27–64.

Mound, L.A. 1977. Species diversity and systematics of some new world leaf litter Thysanoptera (Phlaeothripidae Glyptothripini). *Syst. Ent.* 2: 225–244.

Mound, L.A. and Palmer, J.M. 1982. The generic and tribal classification of spore feeding Thysanoptera (Phlaeothripidae: Idolothripinae). *Bull. Br. Mus.* (*Nat. Hist.*) *Ent. Jr.* 43: 1–174.

Moursi, A.A. 1962. The attractiveness of Carbondioxide and Nitrogen to soil Arthropoda. *Pedobiologia*. 1: 299–302.

Mukharji, S.P. and Singh, J. 1970. Seasonal variations in densities of soil arthropod population in a rose garden at Varanasi. *Pedobiologia*. 10: 442–46.

Mukhopadhyay, A. 1988. Biological notes on two species of big-eyed bugs (Insecta: Hemiptera: Lygaeidae: Geocorinae). *J. Bombay Nat. Hist. Soc.* 85: 298–310.

Murphy, P.W. 1955. The biology of forest soils with special reference to the mesofauna or meiofauna. *J. Soil Sci.* 4: 155–193.

Murphy, P.G. and Lugo, A.E. 1986. Ecology of tropical dry forest. *Ann. Rev. Ecol. Syst.* 17: 67–88.

Nye, P.H. 1961. Organic matter and nutrient cycles under moist tropical forests. *Pl. Soil* 13: 333–345.

Olson, D.M. 1994. The distribution of leaf litter invertebrates along a Neotropical altitudinal gradient. *J. Trop. Ecol.* 10: 129–150.

O'Neill, R.V. 1976. Ecosystem persistence and heterotrophic regulation. *Ecology.* 57: 1244–1253.

Ovington, J.D. 1962. Quantitative ecology and woodland ecosystem concept. *Adv. Ecol. Res.* 1: 103–192.

Parker, G.G. 1983. Throughfall and stemflow in the forest nutrient cycle. *Adv. Ecol. Res.* 13: 57–133.

Parkinson, D., Visser, S. and Whittaker, J.B. 1979. Effects of collembolan grazing on fungal colonisation of leaf litter. *Soil Biol Biochem.* 11: 529–535.

Pascal, J.P. 1980. Wet evergreen forests of the Western Ghats of India. Ecology, Structure, floristic composition and succession, UNESCO.

Pimm, S.L. and Lawton, J.H. 1977. Number of trophic levels in ecological communities. *Nature, Lond.* 268: 329–331.

Pimm, S.L. and Lawton, J.H. 1978. On feeding on more than one trophic level. *Nature, Lond.* 275: 542–544.

Pimm, S.L., Lawton, J.H., and Cohen, J.E. 1991. Food-web pattern and their consequences. *Nature, Lond.* 350: 669–674.

Prabhoo, N.R. 1967. Microgeographic variation in collembola. *Bull. Natl. Inst. Sci. India.* 34: 225–231.

Prabhoo, N.R. 1971. Soil and litter Collembola of South India—1. Arthropoda. *Orient. Insects.* 5: 1–46.

Rai, b. and Srivastava, A.K. 1982. Decomposition of leaf litter in relation to microbial populations and their acitivity in a tropical dry mixed deciduous forest. *Pedobiologa* 24: 151–159.

Rajagopal, D. and Veeresh, G.K. 1981. Foraging activity of the round building termite, *Odontotermes wallonensis* (Wasmann). *J. Soil Biol. Ecol.* 1: 56–64.

Raman, S.S. 1975. Primary production and nutrient cycling in tropical deciduous forest ecosystems. *Trop. Ecol.* 16(2): 140–146.

Reddy, V.M. 1984. Ecology of soil and litter inhabiting arthropods. *Indian Rev. Life Sci.* 4: 149–217.

Reddy, M.V. 1989. Entomometeorology of pine plantations in North-Eastern India. A case study. In: Proceedings of the 11th ISB-congress, Lafayette, 1989. Eds. Driscoll, D. and Box, E.O. Academic Publishing bv. The Hague. The Netherlands, 115–121 pp.

Reddy, M.V. and Venkataiah, B. 1989. Influence of microarthropod abundance and diametric factors on weightloss and mineral contents of Eucalyptus leaf litter during decomposition. *Biol. Fertile Soil* 8: 319–324.

Reddy, M.V. and Venkataiah, B. 1990. Effects of tree plantation of qualitative and quantitative composition of soil arthropods of a semi-arid tropical savannah. *Environ. and Ecol.* 8(1): 361–367.

Reddy, A.S., Kumar, R. and Mathur, S.N. 1978. Studies on the Biology of *Galumna flabellifera* Hammer and its Role as Fungal Spore Carrier. In: *Soil Biology and Ecology in India.* Eds. edwards, C.A. and Veeresh, G.K. UAS Tech. Series 22: 134–143.

Reichle, D.E. 1971. Energy and nutrient metabolism of soil and litter invertebrates. UNESCO: 465–477.

Reichle, D.E., Shanks, M.H. and Crossley, D.A. 1969. Calcium, potassium and sodium content of forest-floor arthropods. *Ann. Ent. Soc. Amer.* 62: 57–62.

Richards, P.W. 1972. *the Tropical Rain Forest: An Ecological Study.* Cambridge Univ. Press, 1952. 450 pp.

Robertson, G.P. and Vitousek, P.V. 1981. Nitrification potentials in primary and secondary succession. *Ecology.* 62: 376–386.

Rodin, L.E. and Bazilevich, N.I. 1967. Production and mineral cycling in terrestrial vegetation. Oliver and Boyd. London, 288 pp.

Roisin, J., Everaerts, C., Pasteels and Bonnard, O. 1990. Caste dependent reactions to solider defensive secretions and chiral alarm / recruitment pheromone in Nasutitermes. *Princeps.* 16(10): 2865–2875.

Rusek, J. 1975. Die bodenbildende Funktion von Collembolan und Acarina. *Pedobiologia* 15: 299–308.

Sabu, K. Thomas, Jeyakumar, A. and Ananthakrishnan, T.N. 1995. Dynamics of insect communities at varying altitudes in the Shola forests of Kodaikanal hills, in relation to the chemical ecology of litter. *Int. J. Ecol. Environ. Sci.* (in press).

Schuster, R. 1985. Determination of fatty acids in mangarima and butter by on column derivatization HPLC application, Hewlett-Packard Publ. 12: 5954–6826.

Seastedt, T.R. 1984. The role of microarthropods in decomposition and mineralisation processes. *Ann. Rev. Entomol.* 29: 25–46.

Seastedt, T.R. and Crossley, D.A. 1980. Effects of microarthropods on the seasonal dynamics of nutrients in forest litter. *Soil Biol. Biochem.* 12: 337–342.

Sharma, G.D. 1967. Bionomics of *Tomocerus Vulgaris* Tulb (Collembola: Entomobryidae) *Proc. R. Enl. London.* (A) 42: 30–34.

Shaw, P.J.A. 1988. A consistent hierarchy in the fungal feeding preferences of the Collembola *Onychiurus armatus. Pedobiologia.* 31: 179–187.

Shindo, H., Ohtoa, S. and Kuwatsurka, S. 1978. Behaviour of phenolic substances in the decaying process of plants. IX: Distribution of phenolic acids in soils of paddy fields and forests. *Soil Sci. Plant Nutr.* 24: 233–243.

Singh, K.P. 1968. Litter production and nutrient turnover in deciduous forests at Varanasi. Proc. Symp. Recent. Adv. Trop. Ecol. Eds. Misra, R. and Gopal, B. *Int. Soc. Trop. Ecol.* Varanasi, 655–665 pp.

Singh, K.P. 1969. Nutrient concentration in leaf litter of ten important tree species of deciduous forests at Varanasi. *Trop. Ecol.* 10: 83–95.

Singh, J.S. and Gupta, S.R. 1977. Plant decomposition and soil respiration in terrestrial ecosystem. *Bot. Rev.* 43: 429–528.

Singh, J. and Singh, U.R. 1975. An ecological study of soil microarthropods from soil and litter of tropical deciduous forest of Varanasi (India). *Trop. Ecol.* 16(2): 81–85.

Sinha, A. and Dayal, R. 1983. Fungal decomposition of teak leaf litter. *Indian Phytopath.* 36: 54–57.

Sinsabaugh, R.L. and Linkins, A.E. 1989. Cellulase mobility in decomposing leaf litter. *Soil Biol. Biochem.* 21: 205–209.

Slansky, F. (Jr.) and Rodriguez, J.G. 1987. *Nutritional Ecology of Insects, Mites, Spiders and related Invertebrates.* John Wiley and Sons, 1016 pp.

Swden, F.J. and Ivarson, K.C. 1959. Decomposition of forest litter. II. Changes in nitrogenous constituents. *Pl. Soil* II: 249–261.

Sowden, F.J. and Ivarson, K.C. 1962. Decomposition of forest Litter. III. Changes in carbohydrate constituents. *Pl. Soil.* 16: 389–400.

Stanton, N.L. 1979. Patterns of species diversity in temperature and tropical litter mites. *Ecology.* 60(2): 295–304.

Stark, N. 1972. Nutrient cycling pathogens and litter fungi. *Bioscience* 22: 355–360.

Stefaniak, O. and Seniezak, S. 1981. The effect of fungal diet on the development of *Oppia nitens* (Acari, Oribatei) and on the microflora of its alimentary tract. *Pedobiologia* 21: 202–210.

Straalen Van, N.M. 1983. Comparative demography of forest floor Collembola populations. *Oikos* 45: 253–265.

Subba Rao, B.K., Dabral, B.G. and Pande, S.K. 1972. Litter Production in Forest Plantations of Chir (*Pinus roxburghii*), Teak (*Tectona grandis*), Sal (*Shorea robusta*) at New Forest, Dehra Dun. In: *Tropical Ecology with an Emphasis on Manic.* Eds. Colley, P.M. and Golley, F.B.

Swift, M.J., Heal, O.W. and Anderson, J.M. 1979. *Decomposition in Terrestrial Ecosystem.* Blackwell Scientific Publication, Oxford, 372 pp.

Szklarz, G.D., Antibus, R.K., Singsabaugh, R.L. and Linkins, A.E. 1991. Production of phenol oxidases and peroxidases leywood rotting fungi. *Mycologia* 81: 234–240.

Tanner, E.V.J. 1981. The decomposition of leaf litter in jamaican montane rain forests. *J. Ecol.* 69: 263–273.

Tate, K.R., Ross, D.J., O'Brien, B.J. and Kelliher, F.M. 1993. Carbon storage and turnover and respiratory activity in the litter and soil of an old growth southern Beach Forest (Nothophagus). *Soil Biol. Biochem.* 25: 1601–1612.

Teuben, A. and Roelfsma, T.A.P.J. 1991. The Role of *Tomocerus minor* and *Philoscia muscorum* in Decomposition of Black Pine litter: A Microcosm Study. In: *Advances in Management and Conservation of Soil Fauna.* Eds. Veeresh, G.K., Rajagopal, D. and Viraktamath, C.A. Oxford and IBH Publishing Co. Pvt. Ltd., New Delhi, 91–98 pp.

Toky, O.P. and Singh, V. 1983. Litter dynamics in short-relation high density tree plantations in an arid region of India. *Agric. Eco. Envir.* 45: 129–145.

Trenbath, B.R. and Fox, L.R. 1976. Insect frass and leaves from *Eucalyptus bicostata* as germination inhibitors. *Aust. Seed Sci. Newslett.* 2: 34–39.

Turner, J.F. and Lambert, M.J. 1986. Nutrition and nutritional relationships of *Pinus radiata. Ann. Rev. Ecol. Syst.* 17: 325–350.

UNESCO/UNEP/FAO. 1978. Tropical forest ecosystems. 683 pp.

Usher, M.B. 169. Some properties of the aggregations of soil arthropods: Collembola. *J. Anim. Ecol.* 28: 607–622.

Usher, M.B. 1970. Seasonal and vertical distribution of a population of soil arthropods: Collembola. *Pedobiologia* 10(5): 224–236.

Usher, M.B. 1976. Aggregation Responses of Soil Arthropods in Relation to the Soil Environment. In: *The Role of Terrestrial and Aquatic Organisms in Decomposition Processes.* Eds. Anderson, J.M. and Macfadyen, A. Blackwells, Oxford, 61–94 pp.

Usher, M.B. 1985. Population and Community Dynamics in the Soil Ecosystem. In: *Ecological Interactions in Soil, Plants, Microbes and Animals.* Special Publcn. Series of the British Ecological Society. No. 4. Eds. Fitter, A.H., Atkinson, D., Read, D.J. and Usher, M.B. Blackwells Scientific Publications. Oxford. 451 pp.

Usher, M.B. and Parr, T.W. 1977. Are there successional changes in arthropod decomposer communities? *J. Environ. Manag.* 5: 151–160.

Usher, M.B., Booth, R.G. and Sparkes, K.E. 1982. A review of progress in understanding the organization of communities of soil arthropods. *Pedobiologia.* 23: 126–144.

Van Veen, J.A. and Frissel, M.J. 1981. Simulation Model of Behaviour of Nitrogen in Soil. In: *Simulation of Nitrogen Behaviour of Soil-Plant systems.* Eds. Frissel, M.J. and Van Veen, J.A. Pudoc. Wageningen, 126–144 pp.

Verhoef, H.A. 1978. An ecological study of the water relations in Collembola. Thesis, Amsterdam (*adopted from* Joose, 1981, *Pedobiologia* 21: 346–356).

Verhoef, H.A. and Selm, A.J. van 1983. Distribution and population dynamics of Collembola in relation to soil moisture. *Holarctic Ecol.* 6: 387–394.

Verhoef, H.A., Prast, J.E. and Verweij. 1988. Relative importance of fungi and algae in the diet and nitrogen nutritious of *Orchesella cinota* (L.) and *Tomocerus minor* (Lubbock) (Collembola). *Funct. Ecol.* 2: 195–201.

Visalakshi, N. 1993. Litterfall, standing crop of litter and their nutrients in two tropical dry evergreen forests in India, *Int. J. Ecol. Env. Sci.* 10: 163–180.

Visser, S. 1985. Role of Soil Invertebrates in Determining the Composition of Soil Microbial Communities. In: *Ecological Interaction in Soil.* Eds. Fitter, A.H., Atkinson, D., Read, D.J. and usher, M.D. *Br. Ecol. Soc.* Blackwell Scientific Publication. Oxford, 297–318 pp.

Visser, J.H. 1986. Host odour perception in Phytophagous insects. *Ann. Rev. Entomol.* 31: 121–144.

Vitousek, P.M. 1982. Nutrient cycling and nutrient use efficiency. *Am. Nat.* 119: 553–572.

Vitousek, P.M. 1984. Litterfall, nutrient cycling and nutrient limitation in tropical forests. *Ecology* 65: 285–298.

vitousek, P.M. and Sanford, R.L. (Jr.) 1986. Nutreint cycling in moist tropical forests. *Ann. Rev. Ecol. Syst.* 17: 137–167.

Vogt, K.A., Grier, C.C. and vogt, D.J. 1986. Production, turnover and nutrient dynamics of above and below ground detritus of world forest. *Adv. Ecol. Res.* 15: 303–378. Academic Press, London.

Waksman, S.A. 1952. *Soil Microbiology.* John Wiley and Sons, Inc. New York.

Wallwork, J.A. 1958. Notes on feeding behaviour of some forest soil Acarina. *Oikos* 9: 260–271.

Wallwork, J.A. 1959. The distribution and dynamics of some forest soil mites. *Ecology* 40: 557–563.

Wallwork, J.A. 1971. The distribution and diversity of soil fauna. London: Academic Press. 355 pp.

Wallwork, J.A. 1983. Oribatids in Forest Ecosystem. *Ann. Rev. Entomol.* 28: 109–130.

Werner, M.R. and Dindal, D.L. 1987. Nutritional Ecology of Soil Arthropods. In: *Nutritional Ecology of Insects, Mites, Spiders and Related Invertebrates.* Eds. Slansky, F. (Jr.) and Rodriguez, J.G. John Wiley and Sons. N.Y., 815–836 pp.

White, C.S. 1994. Monoterpenes their effects on ecosystem nutrient cycling. *J. Chem. Ecol.* 20: 1381–1406.

Whitehead, D.C. 1964. Identification of p-hydroxybenzoic, vanillic, p-coumaric and ferulic acids. *Nature, Lond.* 202: 417–418.

Whitehead, D.C., Dibb, H. and Hartley, R.D. 1981. Extractant pH and release of phenolic compound from soils, plant roots and leaf litter. *Soil Biol. Biochem.* 13: 343–348.

Whitehead, D.C., Dibb, H. and Hartley, R.D. 1982. Phenolic compounds in soils as influenced by the growth of different plant species. *J. Appl. Ecol.* 19: 579–588.

Witkamp, M. 1960. Breakdown of forest litter in relation to environmental factors. *Plant & Soil* 15(4): 295–311.

Witkamp, M. 1963. Microbial populations of leaf litter in relation to environmental conditions and decomposition. *Ecology* 44: 370–377.

Witkamp, M. 1966. Decomposition of leaf litter in relation to environment, microflora and microbial respiration. *Ecology* 47(2): 194–201.

Witkamp, M. 1971. Soils as components of ecosystems. *Ann. Rev. Ecol. Syst.* 2: 85–110.

Witkamp, M. and Drift, J. Van Der. 1961. Breakdown of forest litter in relation to environmental factors. *Plant & Soil* 15: 295–311.

Wood, T.G. 1974. Field investigation on the decomposition of leaves of *Eucalyptus delegatensis* in relation to environmental factors. *Pedobiologia*. 14: 343–347.

Index of Plant Names

Index of Animal Names

Index of General Terms